Water: The Web of Life

WATER
The Web of Life

Cynthia A. Hunt and Robert M. Garrels

W · W · NORTON & COMPANY · INC ·
NEW YORK

1

Library of Congress Cataloging in Publication Data
Hunt, Cynthia, 1941–
    Water: the Web of Life.

    Bibliography: p.
    1. Water. I. Garrels, Robert Minard, 1916–
joint author. II. Title.
GB661.H84   551.4'8   78-152663
ISBN 0-393-06386-0 Cloth Edition
ISBN 0-393-09407-3 Paper Edition

56,925

To Jack Stark, friend and teacher, who keeps our faith in the human race strong, despite evidence that might lead us to a contrary view.

Contents

Preface

THIS BOOK was written for the general reader. We wrote it because we thought that the material showered upon the public by news media, including newspapers, television, general and technical magazines, books, and radio, is so fragmented, has so much emphasis on local situations, and is so highly selected for news value, that a broader perspective was needed. Emphasis is on world water supplies, with the United States used as an example of a typical "developed" country. In addition to providing information on water supplies, the book develops the basic physical principles governing the behavior of water with a minimum of technical material. We have also tried to show how political, technical, and economic factors will be critical in the water supply situation of tomorrow, and how their unpredictability makes it impossible to do more than forecast trends.

A theme that runs throughout the book is our conclusion that whatever the details, the future will require the complete management of water. It is not possible to have streams and lakes restored to their initial purity by recreating the conditions that kept them pure in the beginning. Water must be guarded and purified and controlled and used. There is hope that there will be abundant beautiful rivers and lakes thirty years from now, including many that are in sad shape today, but this can come only from a national or international concerted and expensive effort that must be guided by a central authority whose influence will extend down to the level of the individual citizen.

One of the difficulties in writing a book like this is the inability to give a full discussion of the validity of the numbers that must be used. Some, like the volume of the oceans, are precisely known; others, like the volume of global stream flow, are perhaps within twenty percent of the true value. The supply of underground water is typical of the worst difficulties we faced; most of the numbers we have been forced to use are estimates of the amount of water of a specified compositional range that can be obtained from above a given depth within a particular range of cost per unit of water. To go into the details on which all the estimates are based would be prohibitive; yet the validity of many of our conclusions rests on our having selected good numbers. We hope we have.

Finally, as we put this book together, we encountered the dilemma of power and water at every turn. Water management always requires power; power production is using the earth's resources of coal, oil, and gas at a rate exceeding population growth. Power production causes problems by adding sulfur gases, carbon dioxide, dust, and heat to the environment, or by generating radioactive wastes. Everything we say about the future of water supplies hinges on energy demands and how they will be met.

BERMUDA
AUGUST, 1971

Water: The Web of Life

~~~~~~

CHAPTER ONE

Water, History, and Life

OF ALL the basic materials once thought to exist in superabundance, one of the most important is water. We must have a quart and a half each day to survive; our bodies are 70 percent water; our food cannot grow without it. It is as basic to us as air. Now it too is threatened.

This is the story of water itself—what it is, how it behaves and why, how it cycles from ocean to land and back again—a story of where water is and why it is salty, of why it rains or not. It is the story of the compound that has made earth unique in the solar system. The story of the *Diamond Lens*, in which a beautiful creature is seen to shrivel and die as the drop of water in which she lives evaporates, could become the story of earth. If this precious resource is ruined, life will cease.

Water is a continuous web, a network that unites the whole

world we know. The vast mass of the oceans connects with streams and tributaries deep inside the continents, and water spreads from these through the pores of all rocks and soils. Ocean water evaporates into the atmosphere as a thin gas, thickening where it condenses into clouds. All life forms are really organized and controlled masses of water—tiny drops of watery fluid enclosed in delicate membranes. Only in the densest solids is the web broken, and even there, in rocks and bricks and glass, water cannot be denied forever. It filters slowly in; given time the solids erode and decay. The structures built by man are temporary victories against the spreading of the water web. If we could photograph the world with a film sensitive only to water, the shapes of all objects would remain, but their solidities would be altered and often reversed. Flesh would be dark and bones would be light. Oceans would be black and rocks nearly white.

The continuous web of water is far from static. Solar heat drives moisture from the oceans and continents by evaporation; gravity pulls it down as rain and the water runs back to the sea. The water in a tree is replenished thousands of times as the tree grows. The hundred or so pounds of water in a man's body is replaced seventeen times a year. Water in the atmosphere is but twelve days old; while the oceans' water is renewed by streams every forty thousand years.

Whence came this water that permeates and circulates, and why is it so important? Here we outline the story as we know it now; the record is fragmentary and stretches back five billion years to the beginning of the earth itself. The reconstruction depends upon current theories of the origin of the earth and of its history. Such theories keep changing and are refined and elaborated as more facts about earth history are discovered. The story here, then, is neither final nor immutable.

In the beginning the earth was built of an accumulation of meteoritic particles and of gases drawn from cosmic clouds. Its atmosphere, drawn from the same clouds, contained methane, ammonia, carbon monoxide, and carbon dioxide. Some time not long after this solid earth had grown to full size, the original atmosphere was lost. Such large outer planets as Jupiter and Saturn have atmospheres chiefly of methane, ammonia, hydrogen, and helium. These are the gases found in the cloudy parts of space and have been accumulated by planets of our solar system. The early earth atmosphere probably did not contain hydrogen and helium, for Earth was then so small that its gravitational attraction could not hold such fast-moving molecules captive.

As the earth grew in size it warmed, due partly to the heat engendered by its growth and compaction and partly to decay of radioactive elements such as uranium and thorium in the cosmic dust particles that were swept up by the growing earth. Within a few hundred million years the temperatures became so high that a molten stage was approached, and the originally homogeneous earth began to differentiate into "shells" of core, mantle, and crust. This may have been the time at which the first atmosphere was lost—the surface was hot, the rotation was speeding up, and the sun, also nearly newborn, was bombarding the earth with intense high energy radiation. All these factors would have helped to throw off and strip away the primary gaseous envelope of the primitive atmosphere.

As the earth surface then cooled and the "solar wind" from the sun diminished, a new atmosphere began to form. Such dense materials as iron and nickel sank to make the inner core; such lighter materials as aluminum, sodium, and potassium rose to make the crust. Still lighter substances emerged at the surface as gases, but were held within the strong inner part of the earth's gravitational field.

Water vapor, almost absent in the primeval atmosphere, was the most important gas of the new one. With carbon dioxide, methane, carbon monoxide, hydrogen sulfide, and hydrochloric acid, the water came to the surface and remained as water vapor until the surface grew cool. Hydrogen and helium must have

15

escaped from the interior in large quantities, but the earth could not keep them from slipping off into space. Oxygen was absent.

It is hard to imagine in the modern world, where water is a soothing, drinkable substance and oxygen is a requirement for life, that our environment could have evolved from that deadly acid brew of 4½ billion years ago. But eventually the water vapor managed to condense to form the first ocean, and surface temperatures dropped below boiling. The acid gases began to be neutralized by the solid rocks of the surface until finally the seas were nearly neutral. When hydrochloric acid reacted with rocks, the acid was removed from the seas and from the atmosphere, releasing salts that dissolved in the ocean water. The carbon dioxide and hydrogen sulfide reacted with surface rocks to make various new minerals. The atmosphere was beginning to emerge in its present form.

What of the time scale of these events? The oldest rocks for which there are reliable ages are about 3½ billion years old; the first accretion of the earth has been placed at about 5 billion years, and the earth as we know it now, in terms of having a core, mantle, and crust, at about 4½ billion years.

So we can say that 4 billion years ago or so there was an ocean, that it was salty, and at most slightly acid, that the acid gases had been abstracted from the atmosphere and fixed in rocks or as salts dissolved in the sea. We guess that the ocean of 4 billion years ago may have had a volume of at least one half of that of today. Just how much water vapor and other gases have been released from the earth's interior since that time, or even how much are being released today, we do not know. There may have been continuous but irregular increase in the volume of the ocean.

Whether there were continents projecting above the early ocean, or whether the ocean covered the entire earth, is only to be guessed at. Perhaps the ocean spread around the globe, broken only here and there by a volcano building its cone from the sea floor until it rose above the surface. But, as soon as land appeared, the water cycles began that have carried on until today. When the water that was being ceaselessly evaporated from the sea to rain back upon the sea again fell instead upon the first volcanic island, it washed some loose rock down the slopes. Some of this rock dissolved in the fresh rain water (the salts had been left behind when the water molecules evaporated from the sea surface, speeded by the sun's radiant energy) and was carried back to the sea. As the land struggled to grow the rains attacked it, battling to keep the sea surface unbroken.

The land has won—today the continents occupy thirty percent of the surface of the earth—but the struggle has been nearly an equal one, and there have been times when the sea nearly reclaimed its domain. Geologists continuously search for the remnants of the first land. A generation ago the oldest known rocks had an age of 2 billion years; now some have been found whose age is 3½ billion years.

The most striking aspect of earth as seen from space is its abundant water. Great wreaths of white clouds make complicated patterns above the deep blue of the ocean background. Even when viewed from distances so great that all detail is lost, the bright colors and patterns of earth make striking contrast with the barren, dark, pitted surfaces of Mars and the Moon. We may be surprised, but the more we learn about our dry celestial neighbors the less likely it seems that life can exist upon them.

There are scientists who claim that there are billions of suns and that millions of them must have planets, and that of these millions of planets there must be some just like earth where man could live. But as we learn more about how unusual the conditions on earth really are, and how complex a history was required to get it to its present state, we wonder whether among even millions of planets we could find a duplicate of earth.

When we look at the moon and the earth we can almost imagine that they were created as a pair especially to show the drastic differences that stem from the presence or absence of water. The moon preserves the scars of an eternity of bombardment by fragments of matter from space. Earth would have the same appearance were it not for the smoothing action of water. Only where there is a drastic lack of water and a recent meteorite fall, as at Meteor Crater in Arizona, is there a pit so young that it has not yet been obscured by the weathering action of water. There are some desert areas on earth, such as Death Valley in California, where aridity has immobilized the landscape to give it a moon-like quality—the trails of the gold seekers of 1849 can be followed as if they were made yesterday; pits dug by vanished Indians are unchanged except for a thin layer of dust and sand. There is also a silence when water is absent. No rustle of leaves or pounding of surf, no splash of rain or sounds of animals. All that is left is a silent, sun-baked landscape.

The moon did not surprise us with its lack of water, or even lack of an atmosphere. Even if an ocean were transported to the moon it would not remain; the water would evaporate, and the molecules of vapor would stray off into space. Venus and Mars have been a deep disappointment to those with hopes for other life within attainable distances. The atmosphere of Mars is thin and the moon-like nature of its cratered surface indicates that no sea or rain exists, even though there may be enough water vapor to make frost and ice at its white-capped poles. Venus is deeply shrouded in clouds but we know that its surface is so hot—300° F above boiling—that it has no ocean, nor is water vapor important in its atmosphere.

Mars and Venus and Earth all have densities about five times that of water. We assume that they must have formed by similar processes and from similar materials some five billion years ago. Presumably each is layered into core, mantle, and crust. Because of the differences in their distances from the sun, the energy each receives is not the same, but the earth probably would not change drastically if placed in either of the other's orbits. All three planets have "secondary" atmospheres derived from their interiors. Yet only Earth has an ocean.

Having an ocean might be a special kind of phenomenon. On

Earth, water vapor was the chief gas released from the interior when the "secondary" atmosphere was formed. Because the size and density of Venus are so similar to those of Earth, it is argued that the materials of which it was formed must have been similar. So Venus should have heated and cooled to form an atmosphere and an ocean like those of Earth. What happened? Why is it still hot and where is the water?

We can explain the differences between Earth and Venus in many ways. Let us suppose that they were the same when their crusts were nearly molten, and that their atmospheres were dominated by water vapor and carbon dioxide. What if, as they began to cool, the water vapor on Venus, a little closer to the sun and subjected to more intense radiations, was decomposed into hydrogen and oxygen by intensive ultraviolet light so that the hydrogen escaped and the oxygen reacted with the crust and was fixed there? This would remove water vapor from the Venusian atmosphere, leaving mostly carbon dioxide. Carbon dioxide (with its heat-retaining property) may have made it impossible for Venus to cool, so that the water has stayed eternally bound in its crust.

Whatever the answer to these speculations, there are clearly many ways in which a planet can lose its water. Apparently there is a very particular series of events that results in a planet with liquid water on its surface. Fortunately we have it, and have had it almost from the very beginning of the earth. As we shall see, however, this does not mean that we can afford to be complacent about our good fortune.

We can classify the period between four billion years ago and two billion years ago as the transition from an earth without oxygen to one with oxygen. There seems little doubt that some of the fossils from rocks two billion years old are the remains of simple green plants, so-called photosynthetic organisms that produce oxygen. Today's plants are chiefly photosynthetic. They take in water and carbon dioxide in addition to a few minerals, and make from them the main constituents of their tissues, while releasing oxygen. The oxygen in turn is used by animals,

19

who release carbon dioxide. The present balance is remarkable. The oxygen-producing green plants, which use up carbon dioxide, are exactly offset by the oxygen-using animals which release carbon dioxide. Biologists have developed a sealed container called a microcosm—a tiny world unto itself—into which they put plants and animals in just the right proportions so that the plants produce the oxygen the animals need, and the animals respond by giving off the carbon dioxide the plants require. This tiny sealed up world goes on functioning almost indefinitely. A stable reciprocal population is produced in the balanced microcosm; it neither grows nor diminishes. The moral as applied to the current population increase is too obvious to be further delineated.

We are fairly well convinced that the atmosphere of four billion years ago had no oxygen, and we are fairly sure that there were oxygen-producing organisms two billion years ago. We do not know when the atmosphere became oxygenated. Some of the organisms that developed in the oxygen-free early atmosphere learned how to use carbon dioxide and water to produce organic matter plus oxygen. For them the oxygen could be regarded as a waste product. But its influence on the environment was striking. The oxygen produced began to destroy some of the original gases such as hydrogen sulfide and methane, changing them to substances required for modern life—carbon dioxide and water.

After the deadly hydrogen sulfide and methane were removed, oxygen began to accumulate in the atmosphere. New organisms developed that used oxygen in their body functions, and the composition of the modern atmosphere began to be approached.

Our dating of these changes is poor indeed. Because there are some bacteria-like organisms found in rocks deposited more than three billion years ago, we can say that "life" existed even then. Whether there were oxygen producers at that time is unknown. Even though we are convinced that oxygen production by green plants was taking place two billion years ago, we still do not know how much headway had been made in producing enough oxygen to destroy the deadly hydrogen sulfide and methane in the earth's early atmosphere.

There are hints that oxygen was present two billion years ago but it could not have been in significant amounts. In the famous African gold deposits of Witwatersrand, minerals occur in rocks that were almost surely exposed to the atmosphere at the time of their formation, yet these minerals could not form or persist in the presence of our modern oxygenated atmosphere, nor are they found in contemporary rock deposits. The great iron ores of Michigan, Wisconsin, and Minnesota, dated at little more than two billion years old, contain quantities of iron minerals that are not forming in significant amounts today because there is now too much oxygen present in air and water to allow their growth.

We are not sure there was a thoroughly modern ocean and atmosphere until recently in earth history. But there is agreement that for the last six hundred million years—a vast stretch of time, but only about 15 percent of total Geologic time— conditions have been approximately as they are today. This is to say that the oceans had roughly their present volume and saltiness, continents were present with oceans around them, and the temperatures of land and sea were within today's range. At times the seas partially covered the continents. The resulting greater oceanic area, as the seas crept onto the continents, kept temperatures more uniform than today so that climates were warmer and less differentiated into "belts." Relations between land and sea kept changing with time. Recent studies show that the sea floors are rifting and spreading at rates of fractions of inches per year. Working this movement backward, we might find that two hundred million years ago South America was nestled against Africa!

Except for the continually changing and evolving life forms of both plants and animals, the scene was essentially modern. Water evaporated ceaselessly from the oceans and fell on the continents, collected into streams, and ran back into the sea. Both land and sea were inhabited by varieties of organisms;

there were coral reefs, swimming creatures and burrowing creatures. The various environments were inhabited by organisms performing many of the same kinds of chemical and biological functions as those today, even though they have now been replaced by different species.

The volume of the oceans must have been about the same (even though the area changed as the oceans moved on and off the continents) and the amount of rain received by each square mile of continent about the same as today. Soils developed, water was absorbed into the ground, and lakes and marshes formed. We find in the rock layers deposited during those six hundred million years records of counterparts of all the environments we find today. There were periods of glaciation, recorded by boulder deposits and grooved rock surfaces. There were periods of widespread moderate climates, recorded by extensive coal deposits formed in swamps in which are found the bones of the great herbivorous dinosaurs.

Streams carried dissolved materials, as they do today, and washed along sands and muds to the oceans, as they do today. A picnic and a swim on one of the beaches of four hundred million years ago would be quite pleasant. The waves would not be higher, nor the tides different from today; we could breathe without difficulty, and we would get a moderate sunburn. The sea would be salty and would taste about the same as it does now. So for a very long time the patterns of the earth, if they could have been viewed from a satellite, would have been much the same.

Whether life has evolved within the limits of temperature set by the present earth environment, whether it could tolerate a much greater range, or whether the temperature range is limiting to life, we cannot know. But it is true that living forms inhabit almost every spot on earth, from the frozen tundras of Siberia to the hot springs of Yellowstone Park. Water that is actually boiling is too hot for life. The highest temperature at which living organisms carry on the basic processes of feeding and reproduction is 187° F. The organisms that accomplish this feat are the blue-green algae. It is a tantalizing coincidence that the blue-green algae are among the oldest organisms in the geologic record of life. Continuous existence at temperatures below freezing is, of course, impossible for water-inhabiting forms, but terrestrial animals, like polar bears or penguins, do well in the polar regions.

The transition from inorganic compounds to living organisms that took place sometime before three billion years ago almost certainly required liquid water as the medium, and hence temperatures between freezing and boiling. Another requirement for the presence of life today is maintenance throughout time of earth conditions in which liquid water could exist. One of the advantages of having a great mass of ocean is its tremendous inertia in responding to an external heat change. Water requires more heat to change its temperature than does almost any other common substance, and enormous quantities of heat must be added to change it to steam or subtracted to make it into ice. So it may be that life on earth has survived through conditions of heating and cooling that would have frozen or boiled a lesser ocean and thus destroyed the watery web.

In the development of life on the primitive earth, we envisage two major steps. First is the synthesis of amino acids from the even simpler compounds that came from the interior of the earth, and second is the synthesis of proteins and similar complex molecules from the amino acids. The major elements of living cells are found in the proteins, and the unique character of the proteins depends upon the kinds and arrangements of the amino acids.

For a long time it seemed impossible to believe that the stupendous synthesis of proteins came from such simple starting materials as carbon monoxide (one carbon atom and one oxygen atom), carbon dioxide (one carbon and two oxygens), ammonia (one nitrogen and three hydrogens), methane (one carbon and four hydrogens) and water (two hydrogens and one oxygen— H_2O). Mixtures of the simple compounds would sit together happily for years in the laboratory without any tendency to produce anything more complicated. Books were written about the impossibility of chance resulting in any complex substances that could reproduce themselves, one prime requisite of living creatures.

In 1953, Stanley L. Miller, then at the University of Chicago, performed an experiment that completely destroyed the earlier pessimism and showed that there were indeed reasonable pathways to the development of life under natural conditions. He simulated early earth conditions by mixing the gases methane, ammonia, and carbon monoxide (substances that are poisonous to most living things) in a flask, along with water vapor, and then repeatedly passed an electric spark through it. Brownish substances formed on the walls of the flask. Chemical analysis showed that these were amino acids, the building blocks of proteins. It was immediately obvious that lightning, striking through the early atmosphere, could have been capable of causing the first big step toward living things.

Since then a great variety of experiments have been performed using other sources of energy to cause combination of the simple compounds, and amino acids have been made simply by heating primitive gas mixtures to temperatures quite possible on the early earth.

There was a major problem, however, related to the absence of oxygen in the atmosphere at that time. Without oxygen, the ultraviolet light from the sun would reach the earth's surface with a high intensity. One of the effects of high concentrations of ultraviolet rays would be to destroy amino acids. Many ingenious suggestions were made to obviate this difficulty—that synthesis took place too deep in the oceans for the ultraviolet to penetrate; that rock ledges protected the newly formed compounds, and so on. But none of the suggestions had a convincing ring; if deadly ultraviolet rays were reaching the surface every day, it would be extremely difficult to protect the new compounds long enough for further reactions to build the amino acids into proteins.

Philip Abelson, in 1966, turned the problem of ultraviolet light against itself by showing that sodium cyanide (one atom each of sodium, carbon, and nitrogen) when dissolved in water could actually be changed by ultraviolet light into a mixture of amino acids. Sodium cyanide, by the way, is a deadly poison to man!

We are still a long, long way from being able to tell how the first life was produced by natural processes from the simple substances from which it must have come. On the other hand, we know that amino acids can be built in nature fairly easily, and we also know that scientists are close to achieving the synthesis of life from amino acid building blocks, granting that laboratory conditions are highly controlled and are not those found in nature.

A single cell of a blue-green alga, it must be admitted, is still such a complex chemical system that, despite what has just been said, one still has grave doubts about the natural development of such a system. It is only necessary to watch under the microscope as a cell reproduces itself to have an intuitive feeling that natural chemistry is not sufficient as an explanation. The movements in the cell; the changes in the chromosomes, the governors of inheritance, as they knit together and then split longitudinally into mirror images; the construction of the cell into two new cells, each containing the same chromosomal content; and the growth of the two new cells into duplicates of the original, imply a volition or some vital force, unconscious as it may be.

We stand at the moment, knowing that some of the steps toward first life are easier than we had thought; knowing that synthesis of something very close to life can be done in the laboratory, utilizing only materials that occur naturally; and still being appalled by the gap between the most esoteric laboratory work and understanding of the chemical complexity of the earliest organisms of which there is a record.

Perhaps we underestimate the eons of time available for the evolution of life. After one becomes used to it, one can talk about billions of years quite calmly. But if the earth is 5 billion years old, it means that the sun must have risen and set some 2,000 billion times, and 48,000 billion hours have passed. Between the formation of the oceans and the first life (some 3–3½ billion years ago) there are a billion years for combinations of compounds to take place, for the chance compound to win the battle with all the others.

Whatever the story of the origin of life, water was intimately involved. Cells, whether each is an organism sufficient unto itself or whether they are complexly organized into organs and tissues to form a human being, are mostly water; a complete cell analysis usually shows 60 to 70 percent water.

Even though it is dilute, the cell fluid is distinct from that of the outer environment, whether it is in a single-celled organism floating in the ocean, or in a muscle cell of a dog, bathed in the external watery medium of the other cells of his leg. The difference is caused by the cell membrane, a barrier between the internal and external cell environments. Much work has been done to understand how the membrane keeps the inside solution quite constant, despite changes in the outer environment, and yet manages to permit continuous "communication" between the fluid within and that without. Many scientists feel that the isolation of a watery system by a membrane, thus creating a cell, was a critical step in the development of life. Once a way had been developed to protect an inner chemical system from considerable changes in the outer environment there was a chance for continuity in the chemical composition inside, a stability required for the production of nearly identical offspring when reproduction takes place.

It has already been said that the "flow through" of water in organisms is tremendous. Every three weeks our chemistry is renewed. Although the composition of the inner cell fluids remains almost constant, there is continuous exchange between the inside and the outside. Yet cells are not entirely immune to changes in environment. Cell fluids of most organisms are more dilute than sea water. If man is forced to drink only sea water, the contrast between cell fluids and sea water is too great to tolerate. If one drinks only sea water in large amounts the protective cell mechanism breaks down and death results.

The degree of tolerance between the cell fluids and the external environment has been a major factor in the distribution of the species of animals and plants. Most marine organisms, if placed in fresh water, die; the converse is also true. Biologists have studied salt tolerance in great detail and have concluded that the range of tolerance of most marine organisms is so limited that the salt content of the oceans cannot have changed a great deal in the last 600 million years or there would have been extinction of many marine species that have survived to the present.

On the other hand, there are some creatures that are almost impervious to changes in their external watery environment. Sharks, among the toughest and most successful of all marine animals, have been found in fresh water lakes. The salmon is born in fresh water but has most of its career in the oceans, returning to the rivers to spawn. The opening of the Great Lakes to the ocean via the St. Lawrence Seaway has permitted such normally marine species as the lamprey to invade the fresh water lakes and survive, to the sorrow of the lake trout and white fish.

We can view organisms as structures through which water passes rapidly and which try to maintain a constant internal economy in terms of their cell fluids—upon which depends their survival. They have different ranges of tolerance to the variations of the environment to which they are subjected. But water is all-important to them and they imbibe and excrete it rapidly. They are dynamic entities, changeable within short periods of time by the fluids they are fed, despite the protection afforded them by the cell membranes.

We must remember the strengths and weaknesses of cells when we think of the future. Living things have a remarkable flexibility but they cannot be pushed too far. If we change too greatly the waters on which they depend, they will perish— not all at once, but in a sequence that may be quite upsetting to the continuance of man.

CHAPTER TWO

The Nature of
the Web

POPULATION pressure, growing individual water use, severe local water problems, and a growing consciousness of the far-reaching effects of some pollutants force us to look at the earth's water resources with an eye to measurement.

If we regard the total water in the oceans, in ice, in streams, lakes and rivers, and underground, as an *available* total supply, the water problem seems to disappear. When 71 percent of the earth's surface is covered by oceans averaging two and a half miles deep, how could there be a water shortage? The great Pacific Basin covers half the globe; there were times in the flights of the earth-circling astronauts when they could see almost half the earth at once and yet see only an ocean, dotted here and there with a few tiny islands.

One way to try to visualize the total amount of water is to make it into ice cubes, thirty miles on an edge, each one big enough to cover up most of the city of New York and extending up to the limits of the atmosphere. We would have to melt 12,000 such cubes to make the oceans. Compared to the oceans, the amount of water at any one time in all the other "reservoirs" is tiny. We think of streams and lakes as important but they contain much less than 1 percent of the total amount of water on earth. We hear of the dire consequences of melting all the glaciers, which would raise sea level two hundred feet and destroy hundreds of coastal cities, yet this two hundred feet on a global basis is only 1½ percent of the depth of the oceans. Underground water, water trapped in or moving through the pores of rocks, is the second greatest supply of water and may be as much as 10 percent of the oceanic total. We estimate that all these reservoirs together contain four hundred billion billion gallons of water.

Another way to look at the total supply is to divide it by the number of people who need to use it. A quart and a half per day is required for sheer survival of every human being. There are 3.5 billion people on earth; what is each one's share of the total? The answer is somewhere in the vicinity of 120,000 million gallons of water for every human being now in existence. If each person could have a tank to hold his share of water, the tank would be half a mile on each side. If each person's tank were to be filled from a mixture of sea water, river water, lake water, melted glaciers, and underground water, it would be almost as salty as the ocean itself for 88 percent of all water comes from the ocean basins and would dominate the mixture. So we see that water supply is not a problem of total—at least not for a long time—but of water suitable for various needs and available for use.

Moreover, fresh water has a thousand uses. Water is needed for drinking, for washing, for plumbing, for air-conditioning, for swimming pools, for irrigation, for industry, for boating and fishing and swimming, for pure scenic enjoyment, for transportation, and for the raising of food of any kind. For each use the water requirements differ and for each type of water the supply varies.

Drinking water, if it is to be used continuously with no harmful effects, must fulfill a great number of requirements. It cannot even be perfectly pure, for many of the elements needed for good health come from the dissolved minerals in drinking water. If only pure water is drunk it acts somewhat like a leaching agent and robs the body of essential salts as it passes through our systems. Nor can it contain more than about one tenth of one percent dissolved minerals before it develops a strong, unpleasant taste, and begins to upset digestion.

We are the most complex water treatment factories that can be imagined, taking in enormous quantities of water, extracting and using the elements we need, then expelling them in many ways: through skin, kidneys, intestines, through mouths and noses. When the chemical elements are taken in they are segregated appropriately into the different body fluids and sent on to perform their proper functions. Calcium, strontium, and phosphorus are sent to the bones, potassium is enriched many times within cells, iron goes into red blood cells. The human factory is pretty well geared to a diet of average stream water, but despite the flexibility of the system, a marked increase of almost any element in the water supply causes impairment of the functions of the factory. The dynamic and current nature of the human factory is almost unbelievable. If a person is completely immobilized, he begins to excrete more calcium than is taken in. His bones literally begin to dissolve and do it so fast that the calcium cannot be excreted fast enough to prevent its climbing to dangerous levels in the blood. Astronauts exercise regularly in flight; one of the reasons is to prevent calcium toxemia, as it is called.

"Average" drinking water contains about two hundredths of one percent of dissolved minerals (the same as 200 parts of minerals in a million of water, or 200 ppm). A quart boiled to dryness leaves only about one two-hundredth of an ounce of solid residue behind. This residue is made up mostly of three "salts": common salt (sodium chloride), limestone (calcium carbonate), and gypsum (calcium sulfate), plus some silica (silicon oxide). There are traces of magnesium and potassium salts as well, and miniscule amounts of almost every other element known.

The proportions of the three chief salts differ markedly from place to place; water reflects the compositions of the rocks and soils through or over which it has passed, collecting dissolved minerals as it goes. High sodium and chloride and calcium have little influence on the taste of water, but when the sulfate content goes up water takes on an unpleasant astringent quality. Often when sulfates are high, magnesium is too. Many waters in the arid and semi-arid western states contain a lot of magnesium and sulfate. The combination wreaks havoc with the intestinal tracts of tourists, commonly having a strong laxative action. But the adaptability of the human body is remarkable; natives of the magnesium-sulfate water areas accommodate so well to their drinking water that they not only live happily with it, they may complain that the waters of other places are bland and uninteresting.

It is the flexibility (within limits) of the human body that makes it difficult to say exactly which dissolved minerals and what amount in the water supplies are safe or beneficial. The U.S. Public Health Service sets up standards for the permissible amounts of the various elements in the public water supplies. The task is a continuing one, for careful long-term records are needed to know whether any given substance is good or bad. Because waters differ so much and contain so many different elements, the job of isolating the effects of a given element requires vast amounts of information about the medical histories of the inhabitants of regions served by a particular kind of water.

There are tests for toxic elements like lead and arsenic. There are tests for radioactive substances. There is a growing list of tests for new compounds like insecticides.

One of the surprising bits of information that comes from

reading about the limitations of water composition for the good health of fish, for boiler feed, or for irrigation, is the relative liberality of the standards for drinking water. We seem to be more adaptable than many species of animals and plants in the range of substances we can drink and still survive in good health. We think of ourselves as being delicate; it may be that because of our adaptability we would be among the last survivors in a world of uncontrolled degeneration of water quality.

But no matter what our intake of water we all need a minimum amount for survival. We can go for perhaps eighty days without food, but only ten days without water. A 1 or 2 percent variation in body water is painful; with a loss of 5 percent the skin shrinks, the mouth and tongue become dry and hallucination begins; a 15 percent loss is fatal. We can also have too much water, causing nausea, weakness, mental confusion, disorientation, convulsions and even death.

The body regulates itself quite well and keeps a remarkably constant composition, 70 percent of which is water.

Today, with all kinds of new compounds entering the water supplies, the specification of allowable limits has become almost impossible; new substances are being added faster than their effects can be assessed. In addition, the toxic effect of a compound may be severe when it is tested alone, but when mixed with all the other constituents of drinking water it may be neutralized in one way or another, or it may combine with other substances in such a way as to become more toxic. The poisonous effects of equal amounts of zinc and cadmium are worse in waters also high in calcium and magnesium.

The quality control of water is compounded by the fact that not only does the water have to be safe to drink with respect to its individual components, but also with respect to the rest of the environment. For example, lead limitations set for water take into account other sources of lead—in food and beverages, in the air, in cigarette smoke. The problem is to try to guess how much of this generally unregulated intake can be combined with the intake from water before toxic levels are achieved.

In general, the Health Service bends over backward in its specifications of the maximum tolerable amount of the various elements, but every once in a while comes a rude shock from some unsuspected effect. The kind of complication that is hardest to anticipate is the behavior of various elements during their successive travel through the food chain. Water with a given composition, drunk directly by humans, may be perfectly safe, but one of the substances in the water may be enriched in the tissues of other water users, either plants or animals, then enriched again when one animal preys on another. If the final predator is eaten by humans, toxic concentrations may take their toll. In a sense, then, the water may be satisfactory for direct use but dangerous when considered in terms of its total utilization. DDT, originally at a concentration of only one hundredth of a part per million, has been reported as one thousand parts per million in fish that have eaten fish that have eaten microorganisms from the water originally containing dissolved DDT.

The amounts of different kinds of dissolved mineral matter that can be safely drunk differ widely. The merest trace of lead or arsenic is dangerous because they are cumulative poisons, and, although the amount taken in each day might be harmless, the buildup through the years can produce chronic illness or death. The early symptoms produced by many cumulative poisons— weakness, stomach upset, headache—are so common that they can be attributed to a hundred other causes and are seldom properly recognized. By the time a correct diagnosis is made, internal damage may be irreparable, and there is no way of dispelling lead or arsenic deposited in bones or tissues.

The problems that are beginning to emerge in relation to mercury are typical of the complexities that are encountered today when detailed investigations are made of the occurrence and effects of a toxic element. It has been known for a long, long time that the liquid metal mercury was dangerous. When mercury from a thermometer or barometer is spilled and separates into thousands of tiny droplets, the drops release mercury vapor into the air. If the droplets are left on the floor of a poorly ventilated room, so that the vapor can be inhaled, damage to the liver and kidneys results. Until recently this seemed to be a minor and local peril that could be avoided by a little knowledge and attention to cleaning up mercury spills.

Then two things happened. People began to search for poisonous substances in the environment, and simultaneously a highly sensitive analytical method was developed for mercury detection. It became possible to test for the presence of mercury at hitherto impossibly low levels, and thus to find out just where it occurs and where it is concentrated. During the research on mercury occurrence in the environment, the presence and effects of mercury vapor from liquid mercury droplets were corroborated, and in addition, a new and disturbing relation was discovered. If mercury compounds are discharged into water bodies they accumulate in the sands and muds beneath the water. There, under the influence of microorganisms, mercury compounds can be transformed into an organic mercury compound, methyl mercury. In this organic form the mercury is mobilized and can be utilized and concentrated by marine or fresh water organisms. Moreover, as methyl mercury, mercury becomes far more poisonous than mercury vapor or other inorganic forms of mercury. In man it seeks out the nervous system, where accumulations of no more than a few millionths of total body weight are enough to cause paralysis and death. If methyl mercury enters the body, it takes about seventy days before it is excreted, so a safe daily intake is less than one seventieth of the miniscule amount that causes severe nerve damage.

The current situation with respect to mercury is unclear. There is a great variety of natural sources—minerals in rocks and soils, as well as artificial ones—pesticides, chemical manufacturing processes, burning of coal and oil, old thermometers, barometers, radio and television tubes, antifouling paint. Although the total tonnage of mercury exposed to air and water is small, compared to an element like iron, serious local difficulties can develop where disposal is in places where methyl mercury can be formed and work its way into the food chain.

34

The mercury poisoning that occurred at Minimata Bay in Japan is now a widely known incident. Inorganic mercury was dumped into the bay as waste from a large chemical plant. In 1953 a strange nerve disease appeared among residents of the area. It was traced to methyl mercury-bearing fish from the waters of the bay into which mercury wastes had been dumped. Even though the dumping practice has been stopped, the bottom muds still contain mercury, which can be converted to methyl mercury, released to the water, and so make its way from fish to man.

On the other hand, common salt, found in most waters, can be tolerated well beyond the tenth of one percent that limits the usual natural mixture of dissolved minerals. Even sea water, which contains 35,000 ppm dissolved salt, can be drunk slowly in small amounts, but the upper limit of salt content in water supplies for constant use is about 1,000 ppm, with 500 ppm being preferable. In arid countries much higher salt content is tolerated. In parts of North Africa drinking water with up to 3,000 ppm salts is used.

If a mineral is required for maintenance of an organism and is not provided in adequate amounts in foods, it must be artificially supplemented or be present in water. The classic example is that of iodine. Insufficient iodine causes an enlargement of the thyroid gland called goiter. In midwestern United States the water supplies are low in iodine. Goiter was a widespread disease there until it became common practice to add iodine to table salt. In correcting iodine deficiency, it was simple to do it in a way that permitted each person or family to make a decision about supplementing iodine intake by using iodized salt. An alternative method would have been to increase the iodine in the public water supplies, in which case the consumers would have had no choice but to be medicated. We see in the iodine example the basis for one of the most controversial ethical questions of the day. *Any* treatment of water necessarily adds or removes something and so constitutes a mass medication in the broad sense. The kinds of water treatment that are generally acceptable involve addition or subtraction of substances that are absolutely necessary for general water use. The additions should be harmless, ideally without effect on color, taste, odor, or on any use to which the water is put. This ideal is not attainable, but basic water treatment—aeration, filtration, and chlorination —comes close to fulfilling these goals.

The great fluoride battle still rages; it has not been possible to solve it as easily as the iodine situation. The limits on the concentrations of fluoride that are so low that tooth decay is not inhibited and those that are so high that teeth are damaged are narrow; one half a part per million is too little and two parts per million are too much.

Even these narrow limits are not generally applicable; some people drink more water than others and thus get more fluoride. Especially in the case of children the amount of water drunk is generally directly correlated with the weather. One part per million fluoride may be too much for one person and two parts may be too little for another. To make the situation even worse, most of the direct benefits of fluoridation are restricted to the under fourteen age group. And finally, no entirely satisfactory optional substitute for water in controlling fluoride intake has been developed.

Silica, one of the most common constituents of ordinary rocks, is a major dissolved mineral in many waters, yet it seems to have no physiological effect at all; it can vary sixtyfold from one water to the next without any apparent consequences— beneficial or harmful. Beer, for example, contains four or five times as much silica as most drinking waters but it is obvious from the enormous quantities of beer consumed that silica has no bad effects.

Copper is an essential element for human metabolism. The normal diet may provide only a little more than is required so that an extra supplement in drinking water, although it may impart color, is beneficial. Copper is sometimes added to water to control growth of algae. Iron and manganese, in large quantities, may also cause objectionable color and taste, but they too are necessary for good health.

Cadmium and chromium are not essential elements and can even be highly toxic to the human body. Selenium, zinc, and nitrate also belong to this group. Nitrate is especially dangerous for infants who may drink it in water or milk.

There are many other substances in the water supplies that must be tested and controlled, but to speak of each one and its effects would require chapters if not books. One final consideration is the delivery of a safe water supply, which depends not only on the purity of the source, but on the quality of such equipment as water mains and storage tanks involved in the delivery to the consumer. Maintenance of equipment, which could corrode and release poisonous metals to the water, is vital. Lead pipes were once widely used in plumbing but they have been almost entirely eliminated except to carry corrosive waste waters.

Although it is true that most stream and lake waters in a virgin land could be drunk untreated, the establishment of a small settlement near the water source is sufficient to create local health problems and make water treatment necessary. Of the present 3.5 billion inhabitants of the earth, at least 1 billion regularly drink unsanitary water. Of these, 500 million are continually sick and 10 million die each year. They are the victims of bacteria and viruses and parasites that breed in the digestive tracts of humans and animals, and which, if they get into the water supply, cause diseases like typhoid, cholera, and dysentery. Some of the most important tests performed on drinking water supplies are for bacterial content. Only relatively recently have we learned to control water-borne bacteria and viruses. It was unsafe drinking water that caused massive cholera epidemics in London in the 1800s, yet it was not until the latter part of that century, under the influence of Pasteur's theories, that the idea was abandoned that sickness was caused by impure "vapours" in the air and the necessity of disinfecting drinking water was understood.

Control of microorganisms is by filtration, aeration, and chlorination. Filtration through beds of sand and gravel removes suspended material and some bacteria and dissolved organic particles and clarifies the water; aeration destroys organics and kills many bacteria. If water is thoroughly mixed with air, oxygen "burns" the organic material and the bacteria have little left to live on. Streams with rapids are said to clean themselves in a short distance because of the excellent aeration. Sluggish rivers with much natural organic material, or that have been polluted by organics, release foul-smelling gases as bacteria that live in the absence of oxygen break down the organic material. In water treatment plants the water is sprayed into the air to purify it and to improve its taste.

Chlorination is used around the world as the last crucial stage in removal of microorganisms. Most dangerous species yield to it easily but there are a few bacteria that are extremely resistant. They may ordinarily be absent from a water supply, but if they get in they may cause serious epidemics before they are discovered and the chlorine level and length of treatment are adjusted. It is not feasible to test routinely for all possible pathogenic organisms so the coliform bacillus is generally counted as an indicator of the presence of other bacteria. If coliforms are at a low level then most other harmful bacteria are also found in small numbers. Water containing more than about twenty-five coliforms a quart is unsafe.

Drinking water is also carefully and continuously tested to insure that other undesirable qualities such as taste, color, and odor are kept to a minimum. Iron is one of the worst offenders; it gives water a characteristic taste and can stain clothing at only two ppm. Many well waters with dissolved iron are clear when they come from the faucet but quickly precipitate a yellow iron oxide when mixed with air. The oxide is harmless if drunk but it stains glassware, sinks, cooking utensils and laundry. Waters with dissolved iron often carry a trace of hydrogen sulfide which, in concentrations far less than one ppm, gives a distinct rotten egg odor to the water. All kinds of new tastes and odors entering water supplies today are traceable to organic pollutants of a wide variety.

Man seems to have been molded in his evolution by the water he must drink. If a list is made of the average concentrations of many of the elements in the average river water of the world and beside it is placed a list of the maximun concentrations of those elements that have been established by the U.S. Public Health Service as acceptable for human consumption, we see that injurious concentrations of elements in drinking water are high if the amount naturally present in the rivers is high, and are low if the amount naturally present is low. This is mute testimony that man had to evolve in harmony with the available water. The human machinery is geared to its supplies; anyone born with the ability to tolerate lead in large quantities would derive no survival benefits from his unique talent. In fact, if he had a high lead requirement he would be in trouble, because it is a rare water indeed that contains more than a fraction of a part per million of lead. On the other hand, imagine the difficulties of the person made sick by two or more ppm calcium. He could not use the public water supplies anywhere in the world and would have to set up a private rain collecting system to take care of his needs.

In most communities the water for domestic and municipal uses comes from one common supply which is of drinking quality. It is more convenient to have one supply, maintained at drinking water standards, than supplies tailored to every need. Domestic uses vary greatly from place to place. Water for washing and plumbing and for watering lawns and gardens are some of the uses which help to account for the average domestic consumption of one hundred and fifty gallons per person each day in the United States. This is a very rough estimate and may range from as little as five gallons by a water miser to more than two thousand gallons, depending not only upon the affluence of the household but on the climate as well. To have some idea of how much water an individual uses in a day, it may be of interest to note that the average washing machine uses about twenty-three gallons of water, a dishwasher uses fifteen gallons of water, one flush of the toilet about four gallons, about half a

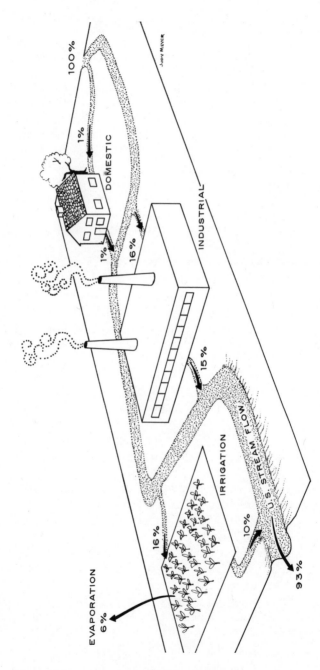

Water use in the United States

quart fills an ice cube tray, four quarts are needed to cook a pound of spaghetti.

Although this domestic water is safe for drinking, it is sometimes treated for "hardness" caused by high mineral content, chiefly calcium and magnesium. The ring around the bathtub comes from the reaction of calcium in hard water with soap. Until the early 1940s the cost of water softening was nearly offset by a saving in the amount of soap otherwise used, but with the advent of the modern detergents this has changed. A familiar phrase used to promote many detergents is, "works even in hard water." However some areas of the country have water so hard that even these detergents are not effective unless the water has been softened. In such areas the municipal supply is softened before it is distributed; otherwise many households install their own water softening units, adding considerable cost per gallon for such a benefit.

Some specialized requirements may be found in home use of water. To prevent corrosion, a steam iron should be filled with distilled water rather than the mineral-laden water from the faucet. Likewise, air-conditioning systems require water with a low corrosive action.

As well as domestic uses, demands on public water supplies may be for recreational facilities. The quality of water for swimming and bathing, while not necessarily as high as that for drinking water, must be safe for human contact. Though the mineral content of bathing water may be of little importance, the bacterial content is.

Another common use of municipal water is fire control, and any water system must be able to deliver a dependable supply in emergencies. Because modern fire equipment uses far more water than the old bucket brigade or horse-drawn pump, water supplies must now be planned with capacities to deliver anywhere from a few hundred to more than twelve thousand gallons a minute, sometimes for many hours.

Domestic and municipal uses (other than industrial) account for less than 10 percent of yearly total water consumption in the

United States. Whereas Parisians of the eighteenth century used only slightly more than one gallon of water daily per person, and today in some parts of the Middle East the average person may use only about three gallons of water daily, most Americans have become accustomed to plenty of clean clothes, frequent baths, green lawns, shining automobiles, and abundant water for drinking and preparing food.

Our personal uses for water are small compared with other uses, though they often seem the most important ones. Even more essential to survival is water for agricultural purposes. Today agriculture accounts for more than 46 percent of total water use in the United States. Again, the requirement is fresh water and in general of drinking quality.

Irrigation is one of man's oldest engineering endeavors. From hand carrying of water to thirsty crops man progressed to digging ditches and building dams to store and divert water. Perhaps the beaver supplied the complicated model for the first man-made dams. Hammurabi, the ancient Babylonian king, described irrigation when he wrote, "I brought the waters and made the desert bloom." We are doing the same thing today, but we have moved to engineering of such magnitude that the giant and far-reaching Feather River project of modern California will eventually have the capacity to deliver two thousand billion gallons of water a year, much of which will benefit agriculture.

The temperature of irrigation water is significant. Warming basins are often provided to bring water to the proper temperature for irrigation; this varies with each crop so that no average figure can be given. When water was first released from Shasta Reservoir for summer irrigation of rice in California, crop damage resulted from the forty-five degree water, sixteen degrees colder than that previously used.

There are approximately four hundred and sixty million acres of cropland in the United States and more than forty-four million of them are irrigated. This 10 percent produces roughly 25 percent of the total value of crop production. Although many areas are already using all the water available to them, irrigation is expected to increase both in the United States and abroad.

Plants as well as animals are biological factories for food production. The water they use affects not only the product but also the factory efficiency. The water they are given must be tolerable to them and safe for human consumption at the end of the food chain. The recommended limit of dissolved solids in water for agricultural uses is about 700 parts per million whereas drinking water for man may have as many as 1,000 ppm dissolved solids.

The amount of water required for agriculture will become greater in the future as population increases. There are already three and a half billion people in the world and by the year 2000 there may well be seven billion. It is predicted that the 46 percent consumption of water by agriculture in the United States will increase to 62.6 percent by 1980. Already Japan, hard pressed by a high population, uses 80 percent of her water to grow rice, and Israel, now using most of her water for agriculture, depends entirely on irrigation for food production.

As an indication of how much water is needed to produce the world's foods, six hundred and fifty thousand gallons of water are required to produce thirty-two bushels of wheat; three hundred gallons of water are needed to produce two and a half pounds of bread; a pound of beef represents twenty-three hundred gallons of water. These figures do not include the amount of water used in processing or transporting food products.

As bacterial spores are perhaps the only vegetable organism having as little as 50 percent water, we can say that plants are mostly water. Lettuce, cucumbers, spinach and asparagus are 95 percent water; tomatoes and carrots are 90 percent water; potatoes are 80 percent water.

Anyone who has weeded a garden has some idea of plants' extensive water collecting system. Most root zones are roughly six feet below the ground surface, although some plants put down roots to much greater depths to obtain water. Trees with roots reaching down forty to eighty feet are not uncommon.

43

Other plants can never reach down far enough to find ground water and their roots may instead spread horizontally beneath the surface. A desert cactus, for instance, may have twenty miles of roots, lying close to the surface so that quickly penetrating rain can be absorbed before it rushes down beyond the root zone. The tiny root hairs may take up many miles in search of water; a single rye plant may have 380 miles of these root hairs.

Another agricultural need for water is for livestock. Although the amounts vary with climate, animals require much more drinking water than humans. Sheep can survive on one and a half gallons a day, while cattle need eight or ten gallons a day and horses as many as fifteen gallons a day. The great herds of animals grazing in the western states alone require staggering amounts of water. One estimate says that there are a hundred million cattle and ten million horses among the United States livestock population. Even with these numbers, the amount of water used for livestock is small in relation to other uses.

Animals can tolerate much more salt in drinking water than can humans. Poultry can survive on water with 3,000 ppm salt and sheep can take water with 10,000 ppm salt. A low salt intake from water is often supplemented with salt blocks. An awareness of animal tolerances of each element in the water supply is as necessary as is this knowledge for control of drinking water for man.

Particularly stringent water quality control is necessary in processing of dairy products. Water for milk processing must be free from all harmful bacteria, yeasts, and molds. And because the flavor of milk is so easily influenced, processing water must be free of tastes. Another strict requirement has been discovered for water used in egg-washing machines, where water containing more than 1 ppm iron may accelerate spoilage.

In 1960, industry and agriculture in the United States each used about one hundred and fifty billion gallons of water a day. Industrial needs are large and varied. It takes, for example, a hundred thousand gallons of water to manufacture an automobile; an average Sunday paper consumes about two hundred

44

and eighty gallons of water in its processing; a ton of steel requires sixty-five thousand gallons of water in the making. The location and size of water supplies determine to a large extent the profit or loss in a business operation. A twenty-two unit apartment house may require more than three thousand gallons of water daily; a laundromat with ten washers may require more than 1,800 gallons of water a day; a car wash that can handle twenty-four cars an hour will need nearly 8,000 gallons of water daily; a large paper mill can easily use more water than a city of fifty thousand people.

Clearly these businesses are not inexpensive to operate in locations with poor water supplies. This water may cost as much as twenty-three cents a thousand gallons from the municipal supply. To operate a laundromat successfully, it may cost an additional three to seven cents per thousand gallons to have "soft" water. Special processing for many industrial water uses adds to the cost.

Not only does the world food supply depend upon water, economic development occurs where there is sufficient water to support, first, the population, and later, the industrial demands. One has to know only a little history to know that civilization flourished with adequate water supplies, and that industrialization has occurred in areas of abundant water for power and for transport of goods. Populations have been forced to move with changing water supplies. Archaeologists have traced the growth and decline of Indian villages in the southwestern United States as water declined in abundance or purity. It has been man's ability to transport enough water from areas of abundant supply to less favored areas, and to otherwise supplement local supplies, that has permitted the development of arid and semiarid regions for agriculture and industry. The technologic demands for further such development will certainly become greater in the next few decades as land with marginal water supplies is pressed into use.

There is much value in the idea that land should be reserved for the use to which it is best suited. Covering good agricultural

45

land with parking lots and highways and tearing down orchards to build more houses takes out of production land with high agricultural yields and eventually forces cultivation of land that does not have adequate water for growing crops, requiring the installation of expensive irrigation systems.

The same dependence on water plagues industry. As water demands increase and supply becomes more scarce, or more polluted and more expensive, industry moves toward abundant and inexpensive water supplies. Industrial water generally must be of at least as high quality as drinking water; in some instances even more restrictive specifications must be met. Some industries buy water from public supplies, then treat it further to meet individual needs. The amount of salt in water used in ultra high pressure steam power plants can be only 1 ppm, while the manufacture of rayon requires water with no more than 100 ppm total dissolved solids. Whiskey can be distilled with water containing 1,000 ppm total solids, so that distilleries may remain longer in polluted areas than steam power plants! Water for cooling, used in numerous industrial processes and accounting for a high percentage of industrial water use, must be noncorrosive and contain extremely low percentages of minerals. Another important factor is temperature; it is obviously important to avoid expensive precooling before water can be used. Some industrial uses for distilled or demineralized water are in leather-finishing processes; the tanning processes sometimes require water with low bacterial content. Each industry has its own special needs for various types of water.

It was once possible to produce power with a wooden paddle wheel and natural stream flow. The water was fresh and clean, and even if it did become polluted the effect on the operation of the mill was small. Today it takes millions of gallons of high quality water to operate boiler-powered electrical plants, and millions of gallons of stream water to produce hydroelectric power. Electricity production accounts for sixty percent of industrial water use.

Why other industries require so much water is often a bit

obscure. The uses for water in food canning will serve as an illustration. Water of drinking quality is used to clean raw foods, which are then transported to various operations in the factory by belts, flumes, and pumping systems (accounting for a major portion of water used). After peeling, fruits and vegetables are rinsed thoroughly in water—especially important if the peeling has been done chemically. Green vegetables are put into hot water or steam to inactivate enzymes and to wilt leafy vegetables to facilitate packing. Very high quality water, free of chlorine, is used for packing, or is used in packing syrups and brines. Containers are sterilized and cooled, both steps using large quantities of water. And finally, water may be used to transport waste materials from the factory.

As another example, the textile industry uses vast quantities of water for washing raw materials, in dyeing and bleaching, and for washing the finished product.

When we survey water use today, we find that industry and agriculture use 96 percent of the total consumed, and that only about 4 percent goes for direct personal uses. One important fact emerges—we tend to think of drinking water as being purer than that for irrigation and industry, whereas the facts are that the tolerance range of humans is greater than that of some plants or of the canning industry. Moreover, as technology becomes more complex, the necessity of treating natural waters for specific uses also grows more involved and costly. As our knowledge of botany, of zoology, of all other aspects of living increases, we find that more and more controls on water composition are desirable.

Study of the water requirements of many species of plants shows that some thrive on high sodium waters; others wilt. With a rapidly expanding world population, it becomes more and more important to reap the maximum production from each acre of fertile land. We find that plants must be fed a diet of water whose composition is tailored to the individual crop.

We are beginning to realize that the natural drinking water supply, even without problems of pollution, or naturally intro-

duced organisms, is not necessarily always the best. Many
natural supplies have enough fluoride to cause bone and tooth
damage; others have enough salt to be damaging to sufferers
from high blood pressure, even though they are satisfactory
for normal healthy persons. At the same time that we are dis-
covering the significance of the many elements occurring naturally in water, we are adding new substances faster than their
effects can be assessed.

Whatever the solution to the ethical and political problems
of control of water composition, some predictions can be made.
"Pure" water, water like that we must drink but with even less
dissolved material, will be required for most industrial and agricultural uses. With increasing utilization of stream and underground water we find that the great reservoir—the oceans—will
be needed more and more. If we are to use the oceans they must
be at least as unsalty as streams. This gives us a clear target for
the future.

What are the requirements today and of the future for this
kind of water? In 1960 the United States used three hundred
billion gallons a day; in 1980 it is estimated that withdrawal of
water will reach six hundred billion gallons a day, and that by
2000 it will be nearly nine hundred billion gallons daily. Water
use grows faster than population. The estimates indicate that
when population is doubled, water use will triple.

If we can hope that by the year 2000 the rest of the world
will have reached a level of affluence more nearly equal to ours,
we must think of water requirements based on per capita numbers. If we use nine hundred billion gallons a day for three hundred million people, we will be operating at a rate of a little
over three thousand gallons per day per person. The best estimate of world population at that time is about seven billion
people. If so, the amount of fresh water required will be close
to twenty thousand billion gallons a day, an amount about
equal to the entire world stream flow.

True, a world in thirty years with uniform international affluence is not very realistic, but even if the growth of affluence is

48

slow, the growth of population is not, so a fresh water demand of this magnitude is certainly foreseeable under any circumstances. The question arises—is that much fresh water available? If not, what can be done about it?

CHAPTER THREE

The World's Water Supply

THE WATER in the oceans, the clouds in the sky, and the ice of the polar seas are all parts of the dynamic solar energy-transfer system that makes the earth run. Our available fresh water comes from this global rain and ice-making machine.

The controls of climate and rainfall depend on just how the solar energy received is distributed to the atmosphere, the oceans, and the continents. The balance is delicate indeed, but tenacious. The earth is just emerging from a great Ice Age, and while there have been other ones in the past, things have never gotten so far out of control as to freeze much of the oceans into ice, or to boil them into the atmosphere as steam.

The earth loses as much energy back to outer space as it receives from the sun (plus a little bit more that escapes through

50

the surface from its internal fires), and has maintained this balance closely for billions of years. Equality of gain and loss has held the average temperature of the surface environment, despite local variations, quite constant. Rocks containing fossils of organisms like those of today that were deposited in the oceans hundreds of millions of years ago show that sea water was about as warm (or cold) then as it is now.

The secret of the ocean thermostat is the ability of water to soak up heat without much temperature change. It takes a lot of heat to raise water to the boiling point and almost ten times as much again to convert it into steam. Even though there may have been great variations in solar energy received at the earth's surface through time, the oceans have kept the temperature constant. When one part of the ocean is heated circulation results, so that it is not possible to boil water at the equator while keeping it frozen at the poles. The faster it is heated, the faster it circulates. The ocean basins are linked together and thus behave as one immense pot—heat part, heat all.

Because it takes so much heat to cause evaporation and so much must be removed to cause freezing, water maintains its liquid state in most natural circumstances. If the energy from the sun were to diminish for a long time, the earth would lose heat to space faster than it would gain heat. Glaciers and ice caps would grow but heat would be generated by the change from water to ice, and, rather than falling below freezing, temperatures would remain at the freezing point until all the water was frozen.

It takes about five times as much heat to cause a temperature change in water equal to the same change in rock. The scorching sidewalk beside the cool puddle is a model of the continents and oceans. It takes little heat to warm rock, and because the rigid concrete transmits its heat but slowly downward, its surface temperature rises rapidly in the sun. Not only does the water of the puddle require more heat per pound for temperature change than does the rock, it also circulates so that the whole puddle must be heated, evaporating eventually into water vapor. For every ounce of water evaporated fifty ounces of rock can be heated ten degrees. The same relation applies to the continents and oceans. In central Asia, summer temperatures may reach 90° F, and in winter may drop to 70° F below zero, while the whole ocean range is only from 32° F to 85° F, and more than 90 percent of all the water is a few degrees above freezing.

If we were going to make another planet for human habitation, with a free choice of materials, we would unhesitatingly demand that the planet have a water ocean to provide maximum safeguards against being scorched or frozen. It takes one unit of heat to warm an ounce of water one degree, it takes almost six hundred units to evaporate it, and eighty units must be removed to freeze it.

Now we get a hint of the importance of having good "communication" between the Atlantic, the Pacific, and the other oceans. The freer the circulation, the smaller the differences in climate from equator to pole; the more restricted the circulation, the greater the climatic contrasts. A planet with land-locked seas, even though they had just as much water as earth's oceans, might have greater polar ice caps and hot seas at the equator.

When sea water evaporates, the salts are almost entirely left behind, and the fresh water vapor moves into the atmosphere. As the vapor rises it cools, condenses, and finally rains back on the sea. If there were no continents the system would be simple enough. For the oceans as a whole, the surface would be lowered about three feet each year by evaporation, but the depth would be restored by condensation and resultant rain. Total evaporation would equal total precipitation.

There is a time lag between evaporation and rain return; the evaporated moisture travels long distances in the atmosphere before it comes back to the oceans. Part of the water evaporated from the oceans moves over the land before it condenses, then it falls as rain or snow on the land, where it may evaporate again, or collect as ice, or sink into the ground, or run off across the surface in rills and brooks to coalesce into rivers that flow back to the sea. Given enough time, all of the water in the oceans will pass down the rivers to the sea. Cleopatra's bath water has run to the sea and mixed throughout the oceans. About 5 percent has already been evaporated and returned to the continents as rain. A few molecules of her bath are present in every tub full of water drawn today.

Yearly precipitation on the continents averages a little less than the three feet that falls on the sea, enough to cover the land to a depth of two and a half feet. Two thirds evaporates into the atmosphere and comes back to the sea as rain, and about one third returns to the oceans via streams. When it first

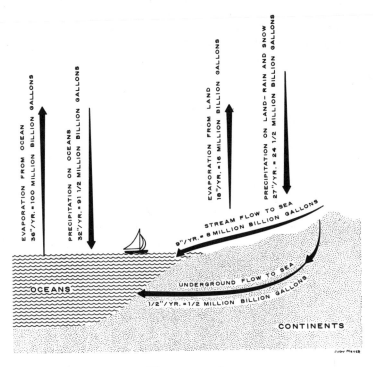

The water cycle

falls and begins to run across the land surface, rain is satisfactory as a water supply for almost every purpose. It can be drunk; it has just enough salts inherited from the sea so that it doesn't have the bad effects of perfectly pure water, yet there is too little salt to have toxic effects. A generation or two ago many homes had a barrel or tank in which rainwater was collected from the roof for drinking and washing, as is still done in Bermuda today.

If we neglect distribution problems engendered by the fact that rain refuses to fall where we want it to, and assume that all that falls could somehow be collected before it evaporated, we come up with a maximum figure for the naturally available water supply. By multiplying the total area of the earth by the average feet of rainfall, we arrive at about forty million gallons of water per year for every person on earth, a total of one hundred and forty million billion gallons.

The sun is very kind in the share of its energy it gives to the water cycle. The amount of energy required to evaporate the one hundred and forty million billion gallons of rainwater is half of the total radiation of the sun received at the earth's surface.

In the United States about one third of the flow of all streams is used at least once, so that we are beginning to use a significant fraction of the major readily available water supply. At the present rate of growth of use and of population we will come pretty close to using an amount of water equal to total stream flow by the year 2000. But *use* does not necessarily consume water. In fact we have not even included under *use* the water that makes electricity at hydroelectric stations, where water turns the turbines and is not changed chemically even though it has performed an important service. Water is consumed only if, as a result of use, it does not return to the stream system, from whence it can be withdrawn again. Eventually, of course, even though it has evaporated or has been taken out of circulation in manufacturing, it will return as rain or snow to land or ocean, and so find its way back some day to the stream supply. It is obvious that some water can be used more than once. It can be withdrawn from streams, used, returned to the stream and used again. The water of the Ohio River is used three times over in its course to the Mississippi, and one of its tributaries is used seven times. If there were no losses by evaporation when water was withdrawn, and if everyone returned the used water in the condition in which he found it, the present supply would be sufficient forever. If it were not for evaporation losses, treatment of used water, perhaps polluted by minor contaminants but not salty, could almost solve our problems. In theory, every house could have a tank and all the water from dishwashing and bathing could go through a purification plant and be returned to the tank. But there would be a net loss; water is drunk, used for gardening, evaporates from the swimming pool, is lost as steam from the teakettle.

The best estimates are that water consumption today is only about one third of water withdrawal. Most of the consumption is attributed to evaporation from irrigation. Little of the evaporation loss is directly from water that is spread on the fields, but is due to evapotranspiration of growing plants. An acre of corn can withdraw three thousand gallons of water from the soil each day and send most of it out into the atmosphere through the leaves as water vapor. Of course what has evaporated is available again at some future time. The general water cycle is one of income, storage, and outgo, much like a checking account.

There are quite reliable figures for the water loss during irrigation; only 40 percent of all the water used is returned to streams. This is in contrast to municipal use, in which 90 percent is returned, or industrial use, in which 90 percent is returned. The problem of how much water is *lost* is still a tricky one because no one knows how much of the water evaporated from a lettuce field in California ends up in the Mississippi River after it has condensed and fallen as rain in eastern Iowa.

But most predictions of the growth of irrigation indicate that it will soon begin to lower significantly the volume of the stream flow of the whole earth.

The degree of reuse of water is determined to a large extent by the change in composition that takes place when water is withdrawn, used, and returned. Water used for baths or showers is little altered. It has had a little dirt and soap added, but it is otherwise unchanged. If bath water could be isolated from the household waste water and treated it would cost little to restore it to its original condition. In contrast, consider the water that emerges from kitchens that have various types of disposal units that grind garbage into sludge that is flushed into drains. This water is loaded with a variety of groundup organic materials that must be oxidized away or settled out before the water returns to potable quality.

Industry poses the same variety of problems. Water used as a coolant in steel making returns to its source unchanged, except for added heat. However, water used in some of the chemical treatments of iron and steel becomes rich in chromium, sulfuric acid, or other chemicals. To clean up cooling water costs little; to clean up chemically altered water costs a lot. There is no simple solution to the reuse problem.

To return water to streams in a condition similar to that in which it was taken requires separation and special treatment of the water, according to the specific uses to which it was put. The cleaning treatments cost money, so that the problem is one of economics, rather than merely of supply or methods for cleaning.

In addition to the constant supply of water that comes from rain we must consider the amounts stored at present in lakes, in glacier ice, and underground. All of these natural reservoirs can be "mined." By mining we mean using a supply for which rate of withdrawal exceeds rate of replacement. We cannot view mining of water reservoirs as a practical approach in the long run. How long could we exist if we did not use the normal daily atmospheric supply, but relied entirely on melting glaciers, or on draining the Great Lakes to supply water to Arizona? If we do not insist on a perpetual water supply and decide to take water only from glaciers, lakes, and underground, how long could man survive before all these reserves were depleted?

The water locked in ice, mostly in Antarctica and in Greenland, is about 1.5 percent of the water in the oceans. But the oceans are so vast that 1.5 percent of them is still a great deal of water—about 5 billion billion gallons. If we could melt the ice and run it into water mains instead of drawing from rivers, there would be enough water, before Antarctica and Greenland were stripped to bare ground, to take care of needs for thousands of years at the present withdrawal rate. The barrier to using meltwater at the present time is economic. Energy is required to melt ice and there would have to be some way of moving the water to places where it would be used, a further addition to the cost.

The idea of using the ice caps is not so far from being economically reasonable as one might suppose. A fascinating suggestion was made by Professor John Isaacs of Scripps Institution of Oceanography—to tow an iceberg from Antarctica, and collect the fresh water as it melted offshore of water-starved southern California. Isaacs estimated that two battleships would be required for towing the ice to San Diego. The project has not yet gone forward, but it is not because the towing costs would be too high, nor because the iceberg would melt en route, nor because the amount of water gained would not be a significant addition to the local supply. Instead the difficulty is the lack of a way to collect the fresh water as the iceberg melts. That particular problem is a difficult one but apparently not impossible to solve, because the meltwater from an iceberg is less dense than salty sea water so the fresh water would tend to float on the sea water surrounding the iceberg and might be collected before the two mingled.

A good sized iceberg may rise one hundred and fifty feet above the sea with seven hundred feet or more beneath the surface. A cubic iceberg eight hundred feet on an edge would melt to about three and a half billion gallons of water, or about 1 percent of the entire daily United States demand. The population of San Diego County is just about 1 percent of United States population. Therefore, if all the water could be collected from an iceberg, San Diego County needs could be served but an iceberg every day would be needed to do the trick! At any rate, it is comforting to know that the great glaciers and icebergs exist, and that an increasingly clever technology may some day solve the cost problem.

A scheme akin to Isaacs' in imaginativeness, but further from possible achievement, has been proposed in which large iceballs would be sent by pipeline from Antarctica to Australia, the heat of friction melting them en route so that they would arrive on Australian croplands as ready-to-use irrigation water.

The amount of usable water in lakes is only about 0.3 percent of that contained in glaciers, but lakes are ordinarily much more conveniently located. Their water can be used without supplying, by one means or another, the tremendous amounts of energy required to melt large masses of ice. The lakes of the world contain about as much water as flows down all the rivers in a year. Lakes are delicate systems; if we began to use them faster than they are replenished, the effects would be so complex and far-reaching that it is almost impossible to estimate what all the consequences would be. If we started to draw water out of Lake Erie, for example, at a tremendous rate, the normal outward flow down the Niagara River would be halted but the inflow to the lake would not be increased. If lake level were lowered 100 feet, water could no longer circulate as it now does. Niagara Falls, between Lake Erie and Lake Ontario, would dry up and Lake Ontario would be cut off from its water supply. We find that use of lakes turns out to be much like use of rivers. We can take from lakes part of what goes into them, or even all that goes in, if we return it; but when we begin mining them, using them beyond the level at which they are restored, the results will be drastic and complicated. Temperatures will change, the fish population will be affected, new kinds of plants

58

The World's Water Supply

will grow, many of the present ones will die, local weather will be affected, even recreation on lakes will be changed. It is not impossible, perhaps, to manage lakes while they are being depleted, though it would be extremely difficult and costly. Lakes can be used as reservoirs to tide us over during years of drought, but long term withdrawals from lakes probably cannot exceed the amount of water flowing in and out of them. Thus they are really only a part of the river system, when long-term water supply is assayed.

The major source of readily available water not being exploited is the underground supply. The total amount in the pores of rocks may be as much as 20 percent of the whole oceanic volume, but only a small fraction can be recovered. The mineable underground water is about a third as much as is locked in ice, and half exists within twenty-five hundred feet of the surface, where it can be fairly easily drilled into and pumped out. The other half is not only expensive to find and retrieve but tends to have too great a salt content for most uses. In the United States the easily reached high quality supply would last five thousand years if we switched to a policy of getting all our water from underground storage. Again we see that the fresh water supply, as a total available amount, is not a serious current problem. In the case of groundwater the difficulty, like that of ice but not so severe, is in its geographic distribution. The eastern and central states have abundant supplies, whereas the western states are in serious trouble. In the Los Angeles area the problem is acute; there is dense population, low and irregular rainfall, and limited groundwater storage.

Where there is adequate rainfall, evenly spread throughout the year, there is plenty of fresh water in all reservoirs—streams, lakes, and underground—but when rainfall drops below fifteen to twenty inches a year, distressing things happen at a pace that accelerates as the rainfall diminishes; the percentage of evaporation increases and water that enters the soil no longer percolates through the ground to swell the streams. Instead it evaporates back up to the surface, carrying with it salts dissolved from the

59

soil. The moisture content of soils in these areas is too small to support agriculture, and in some cases the evaporating water leaves the soil too salty for agriculture. In such areas, water taken from the underground reservoirs, filled perhaps thousands of years ago, is not replenished.

The balance is precarious. It recalls the observation by Mr. Micawber in Dickens' *David Copperfield* that "if a man had twenty pounds a year for his income, and spent nineteen pounds nineteen shillings and sixpence, he would be happy, but that if he spent twenty pounds one he would be miserable." Where evaporation exceeds precipitation, as it does throughout much of the western United States, the result is misery indeed —especially in those areas that have good soils or pleasant climates which attract a large population, with heavy water demands for irrigation, homes, and industry. In these places groundwater is mined at such a high rate that the entire supply can be used in a few years if care is not taken to prevent depletion.

Another question arises: by building dams and creating artificial reservoirs, can water be stored today for perennial future use? The answer is a qualified no. When a river is dammed the reservoir fills with water, then flow resumes as before. There is stored whatever number of years of river flow is required to fill the reservoir, ordinarily only a few years. The great advantage of dams and reservoirs is in creating a method of equalizing and controlling stream flow. Flood waters are collected, then dispersed during dry seasons. Like natural lakes, reservoirs can compensate for rather prolonged droughts. But the consequences of tampering with nature enter here, and they are far-reaching. If artificial reservoirs trap the sand and silt carried in streams, the sand supply downstream is cut off. A beach fed by stream sand may disappear if an upstream dam is constructed. For example, famous Waikiki beach in Hawaii is artificially supplied with sand today. The river that fed it has had its flow cut off from the sea.

The silting problem is always present. In a matter of a few

to hundreds of years every artificial dam has a deposit of sand and silt built up behind it, thus lowering the storage capacity of the lake behind the dam, increasing pressure on the dam (possibly to dangerous levels), perhaps making the lake so turbid that fish cannot survive, and so on. A dilemma confronts every proposal for dam construction: it is necessary to weigh the resulting benefits (flood control, power production, water reserve, recreational facilities) against the drawbacks (silting, downstream changes in water and sand, danger of dam bursting).

To sum up the present world situation with respect to fresh water supplies of the required quality, we see something like this: with no regard to cost, there is plenty of fresh water available for a long time. The sun runs a distillation factory of fabulous proportions and puts enough good water into streams to take care of withdrawals for a generation or two to come. Even in a country like the United States where population is increasing steadily and water use is accelerating faster than population growth, there is enough "new" water in streams each year to carry us for a long time, if the problems of distribution of that water can be solved. Also, despite the fact that the day will come, in a generation or two, when the withdrawal of water from streams will equal the total stream flow, the potential exists today and will certainly be enhanced tomorrow of water treatment that will permit multiple reuse of water withdrawn.

The major natural water problem in the United States and over the rest of the earth is the irregular distribution of fresh water, in ice, streams, lakes, and underground. Most water shortages could be solved by cheap transport of water from one place to another. The stream flow of Canada's waters could support all the water-deficient regions of the United States and Mexico for years to come. The Mackenzie River of Alaska alone, diverted southward, could fulfill the California requirements. The Amazon, carrying 15 to 20 percent of all the stream water on earth, wastes its pristine waters on half a continent that can use almost none of it. Unfortunately there is no water transport now cheap enough to make redistribution of the Amazon possible.

The other great problem is pollution. The whole supply picture can be upset by the addition of a pollutant that makes an entire stream or lake unsuitable for use. Even if withdrawals from streams are now no more than a third of total flow, if we return water that is polluted and which in turn pollutes the rest of the stream beyond use, we have accomplished the equivalent of consuming all the stream water. The result is the same as taking all the water of every stream and using it for irrigation in such a way that the entire amount is evaporated by plants. Even in that case there would be some benefit to food supply. But a serious pollutant cannot be measured in terms of the water it first spoils; it must be measured by its far-reaching effect in making other waters unusable. We can say that the pollution rate today is such that it increases water consumption. If unchecked, this will lead to a situation in which consumption will far exceed withdrawal and will encroach seriously upon the total fresh water supply.

The water budget for the United States can be surveyed using the present withdrawal rate of 450 billion gallons per day as a yardstick. Total stream flow is about 1,100 billion gallons per day, and, since most water withdrawal is from streams, this means that about one third of all the water that flows from them to the sea has been removed and returned at some point between their sources and outlets. A little less than a third of the water withdrawn is consumed: about 10 percent of the total stream flow is not returned to the streams.

It is estimated that by the year 2000, 20 to 25 percent of the stream flow will be consumed, so that an amount of water close to the stream flow total will be withdrawn somewhere across the nation. For some streams this means that their water will be removed and returned five or ten times; for others, only a small fraction of their flow will be moved out and back in. Ironically, consumption is chiefly the result of irrigation; irrigation is required in areas of low rainfall, and areas of low rainfall have few streams. In the arid parts of the United States, *all* stream water will be consumed and streams will shrink to nothing from source to mouth.

At present only 2 percent of our water comes from wells. In the southwest, with its inadequate surface flow, underground supplies have been tapped heavily and in many areas have been exhausted with no hope of replacement.

The city of Tucson, which now blooms with some 300,000 people in the desert of southern Arizona, plans to expand its population and supply water needs with a coordinated program that includes bringing snow water down from the mountains, finding new underground supplies, and attempting to lessen evaporation from reservoirs and from irrigated fields. Even with this program of conservation and development the growth of Tucson is clearly limited unless a major new supply can be transported across the desert, and the scale of irrigation decreases as the city grows.

Not much can be done about increasing the supplies in the arid areas. There probably are no methods available or foreseeable that can increase the rainfall in the deserts, although the statement must be qualified by recognition of the always unpredictable ingenuity of man. The deserts and semiarid deserts lie where they do because of worldwide weather patterns, or because of the presence of mountains that steal moisture on their windward sides as air is forced up and over. A recent report from the atmospheric scientists gives little hope of regional modification of climate by man. Present attempts to water the deserts are the direct approach—bringing water from rainy areas in ditches, pipes, or tunnels.

In densely populated areas, even if they are usually well-watered, the supply problem can become amazingly difficult if there is a prolonged and unexpected drought. Weather records are not ancient enough to show just how much precipitation can vary on a long-term basis. Even good records for a hundred years or more may not be enough to predict the possible extreme ranges of climate. When a four-year drought hit the

New York–Pennsylvania–Delaware–New Jersey area in the 1960s, severe water shortages occurred. The shortages were caused to a large extent by the limits of reservoir planning that had been keyed to probable needs and supply, but were not adaptable to the extreme shortage that occurred. There was plenty of water in the Hudson River to keep New York City supplied, but there was no quick way to set up desalination and water treatment plants to make the water drinkable or even ways to take it from the river and dispense it.

We keep stressing the dynamic nature of the web of water and therein lies the hope for the future of man. Even if we succeed in spoiling all the lakes and underground supplies, even if we find ourselves in such a state that we have polluted all the streams, we can still look forward to a fresh supply of water tomorrow from evaporation of the oceans. That supply is our final great resource.

If all pollution were stopped tomorrow, how long would it take to renew the multiple sources with clean water? What if we were immediately to stop puffing sulfur dioxide and lead and nitrogen oxides into the atmosphere, stop putting DDT and phenols into streams, or phosphates into lakes?

It would take about twelve days to replace all the water in the atmosphere, and perhaps several years before most of the impurities were "rained" out. Compared to the total fresh water in ice, lakes, or underground, the amount in the atmosphere is trivial, but it has a fast turnover. Because streams store about ten times as much water as does the atmosphere, and because they are like the atmosphere in their rapid rate of turnover, the rivers would be flushed out quickly if we could prevent further pollution.

As storage capacity increases, so does renewal time. Lakes store ten times as much water as streams, but still constitute only about 0.3 percent of the total stored fresh water. If the water becomes polluted it will take as much time or more to exchange polluted water for fresh as to fill the lake originally. Lake Michigan, if it became heavily polluted, might require a hundred years or more before its waters were exchanged by clean supply and no one can predict the residual effects of having changed the bottom sediments. As we shall see later, streams can be cleaned up much more easily than lakes. If the water in a stream is changed, many of the effects of the preceding bad water tend to disappear quickly. But a change in the water of a lake is often accompanied by more drastic effect on its plants and animals and on the bottom sediments stored in the lake basin. These effects may require longer to reverse than the time taken simply to replace old water with new. In many cases the changes may not be entirely or even mostly reversible. Because the water cycle does not function in isolation from the rest of the environment, the interlocking effects of many systems must be known if, in fact, lakes and streams are to be cleaned up.

Underground water, the greatest storehouse of available fresh water, represents an accumulation of tens or hundreds of thousands of years, and its renewal time is just as long. The sandstones that are yielding, from deep boreholes, great quantities of high quality water for the city of Chicago, got that water from rain that fell on the Great Plains far to the west a million years ago, then trickled slowly through the rocks at rates of a few feet or a few inches a year. The supply is large, and even with mining, the amounts in the middle west and along the eastern and Gulf coasts of the United States constitute a tremendous, if irreplaceable, resource for thousands of years at present rate of withdrawal. It goes without saying that contamination of underground waters, unless they are very shallow, is difficult and often impossible to correct. On the brighter side, waters moving underground through the pores of rocks tend to be cleaned as the rocks through which they pass filter out contaminants. Thus if pollutants get into the groundwaters, many undesirable substances may be removed, and the moving water again becomes usable.

There is a distinction between underground water that is "permanent" or stored and that shallow water that is really a part of the stream system and acts to keep streams flowing evenly throughout the year. When it rains, some of the water runs directly down the slopes and joins the streams, while another part sinks into the ground. Beneath the ground this water "percolates" through the soil, most of it eventually getting into the streams, although, because of its slow travel, it enters the stream long after the rain has ceased.

In temperate humid areas like England, or the eastern United States, the pores of the soil a few feet below the surface are always filled with water. When it rains, this saturated zone or "water table" rises toward the surface. Between rains it sinks down again as the soil water drains into the streams. Without this pumping action, streams would flow only during and shortly after rains, as is the case in drier areas. We regard this shallow subsurface water as part of the stream system. Even if it is tapped by shallow wells, we do not consider that it is "mined" by this removal, for the water is renewable. Only a small portion of this "soil water" manages to sink deep enough to enter the rocks themselves to become part of the stored underground supply.

The tremendous significance of climate begins to emerge clearly when we contrast what happens in places where precipitation exceeds evaporation with those areas where the reverse is true. When evaporation equals precipitation there is no water left over to percolate through the soil to keep rivers flowing between rains, and there is no regular addition to either deep or shallow groundwater. When it rains, water seeps into the ground but is evaporated out again, carrying dissolved salts that collect in the upper soil layers. In such areas as the Los Angeles basin, water must be supplied by storing rain when it falls and runs off, by transporting water from elsewhere, or by tapping the waters deep down in the rocks. The shallow groundwaters are too salty to be useful. The citrus trees that were so important in the early days of the growth of the Los Angeles region required large amounts of irrigation water, wells had to be drilled deeper and deeper, the water-bearing rocks had their supply depleted, water costs rose. There also followed a settling of the land and in places salt water from the sea replaced the fresh water drawn from wells.

In some arid countries even the deep rock waters may be too salty for use because they were concentrated by evaporation before they trickled down into the ground. Western United States was much wetter during the Great Ice Age, when a great deal of the stored deep water managed to accumulate, but the excess of precipitation over evaporation, even at that time, was not great enough to make much of the country humid.

Deep underground water is thus regarded as a tremendous resource, despite the fact that it usually is located where it is not needed, and scarce or salty where it could be most useful. Nonetheless, it has been an important factor in permitting development of much of the western United States, although its further exploitation is limited.

The situation in the dry areas of the United States is fairly typical of that in semiarid or desert areas elsewhere. There have been some spectacular and important discoveries of groundwater resources, such as in the Nubian sandstone which underlies large sections of the Sahara desert and which is being exploited with deep wells. Use of the waters of this sandstone layer will permit an increase in the population of the Sahara for a long time, despite the fact that much of it will have to be desalted before use. Unfortunately the history of arid areas that depend on groundwater for the maintenance of their populations has been short. In northern Texas an area was developed that had water at hand from deep wells, some seven thousand billion gallons of reserve. In fifty-eight years the population of the area more than tripled and 10 percent of the original water supply is gone. With projected population growth and water use, the remaining supply will last only twenty-five years. The half-million people in that area will have to transport water from elsewhere, change their economy, or move away.

Israel is a fine example of what is to be expected in many more dry areas. The Israelis have been leaders in research on efficient use of water and in developing and conserving water resources in every conceivable way. But it is estimated that unless they can find forty million more gallons for daily use, more than one and a half million Israelis will face severe shortage of irrigation water by 1980.

It should be clear that the adequacy of the water supply cannot be assessed in terms of national or world averages. There tends to be either feast or famine, varying regionally. The best answer to demands of populations spread into poorly watered areas is often an engineering problem insoluble with present technology. Most of the continent of Australia is desert; at the moment one must look with little more than despair at the possibility of heavily populating Australia.

What then should we look for in the future? First of all, in the developed countries we will see the densely populated areas that are now suffering or beginning to suffer from water shortages having to stabilize their populations and hang on to present levels of water use by all kinds of techniques. Water will be brought by pipes and canals great distances. Water will be reused many times, this being made possible by increasingly expensive methods. Pollution will be controlled by limitations on additions to waters used and by water treatment. Agriculture will become more and more sophisticated; better utilization of irrigation water will be practiced, perhaps crops will be adjusted to the nature of the water, with salt-tolerant crops downstream of salt sensitive ones. Watersheds will be planted with vegetation that will produce the most usable flow of water. Practices to reduce evaporation during irrigation and from reservoirs will be introduced. More water will be stored, with better utilization of the rain that does fall, and flood waters will not be permitted to rush off to the sea. Deeper and deeper wells will be drilled, gradually using all the deep groundwater. Desalination of sea water will save critical areas. There will be much research on trying to control rainfall, but probably it will produce local rather than regional or national benefits.

The overall picture is a discouraging one for the proponent of individual rights. Maintenance of communities will demand a degree of planning of water utilization that will involve every town, every city, and whole groups of states. We already know that the use of the waters of the Colorado River in Utah affects the development of California, and cooperative agreements have been made by California, Arizona, Nevada, and Utah to control the development of the Colorado River. Mexico, too, is involved in the use of Colorado River water. As population and water use grow, cooperation will have to extend all the way down to the individual level. The time is probably not far off when worldwide cooperation will be mandatory.

At the moment it looks as though population growth will take place, in the United States and the world as a whole, in the well-watered areas. The state of Georgia has had advertisements in one of the magazines of the chemical industry publicizing its water resources. Georgia has fifty inches of rain a year and a stream flow of 39 billion gallons a day—nearly 10 percent of the national daily withdrawal. Perhaps Georgia will be among the areas of rapid population growth.

The Midwest is also well watered, especially in the northern part, where the Great Lakes, the thousands of smaller lakes of glacial origin, many streams, and abundant underground supplies have been important in supporting the dense populations of northern Ohio, Indiana, southern Michigan, northern Illinois, and southern Wisconsin. Despite the current problems of pollution of the Great Lakes, population is increasing.

Moving west of the Mississippi, average rainfall begins to diminish, and the excess of rainfall over evaporation becomes smaller and smaller, until farming without irrigation becomes difficult. The 20-inch rainfall line, which roughly marks the boundary between enough water and too little, swings back and forth through the years in an irregular cyclic way. In the devastating droughts of the thirties, the 20-inch line was close to the Mississippi River; since then it has stayed far to the west. The difference between the countryside around Sioux City, in western Iowa, during the thirties and today is almost unbelievable. Then, watering reservoirs dried up, cattle were dying on the prairie, corn shriveled and turned brown before it had reached a foot in height, the cropless soil was carried away by the wind. People sat looking hopelessly out across the yellowed plains. Today, the countryside is lush, fat cattle graze, and cities are busy and prosperous.

Farming on the edge of the 20-inch rainfall line is a gamble. Fortunes have been made by men who studied climatic cycles, then bought land cheaply when the drought line was far to the east, cashed in when it came back westward, and managed to get out again before another eastward shift.

It is in the use of the western lands that the complexities of trying to determine what constitutes an adequate water supply is most easily seen. Hit-and-run farming will not work in the long run; to stabilize an area that sometimes has enough rain for corn, sometimes enough for wheat, and sometimes too little for almost anything, requires farmers with foresight and training. They must be ready to switch crops as rainfall lessens. They must plan water storage to provide for livestock and silage. Perhaps they can switch to irrigation if they have wells and adequate pumpage. If they are good managers, they can store part of the excess rainfall in good years, and portion it out with maximum efficiency in dry years. Practices of this kind can change an area from one classified as water deficient into one of adequate water. It certainly is not easy to do, yet it gives a hint of how much more difficult the problem is for industry and for cities.

Perhaps the fabulous increase of population in California and Arizona is nearly at an end, not least of all because of their water supplies.

We see in the next ten or twenty years two major trends: transportation of water from place to place on a currently unbelievable scale and simultaneous adjustment of population growth to water supply. In part there will be conventional and new methods to bring water to the starved areas, and at the same time long-range planning will help to change water use in the underprivileged areas toward domestic and industrial use and reuse, with the progressive abolition of irrigation.

CHAPTER FOUR

Water and Atmosphere

WE SEE nature in a simple physical sense as atoms or space. When the atoms are close together and maintain their relative positions, we speak of solids; when they are close together but free to change their relative positions, they are in the liquid state; when they are far apart and can move so freely that they occupy uniformly any container into which they are placed, they constitute gases. Solids keep their shapes; liquids take the shape of their containers, but do not have to fill them; gases are infinitely adaptable to their surroundings, and swell or shrink to fit their confines. Water exists in all three states in our ordinary experience—from sink to ice tray to stove is but a step.

The attractive electrical forces between the molecules dominate the properties of ice and liquid water, keeping their volumes close to those expected from a collection of spheres almost touching each other. The molecules are in rapid motion, vibrating in solids and rolling over each other in liquids, but they are controlled by their attraction to each other. In gases, the molecules finally are free; all is darting motion and collision, their attractions to each other have no chance to be effective during the infinitely brief moments of collision and rebound. Thus the major change in behavior comes when solid or liquid is changed to gas, as the molecules break their bonds and become free-flying individuals, influenced only by their brief encounters.

A glass of water sitting on a table is not very exciting to our ordinary view, but we forget the never-resting chaotic world we would see if we could magnify the glass ten million times. There would be ten trillion trillion water molecules rolling and bumping within the glass, some going slowly, some fast. In places they would be packed tightly together, elsewhere there would be almost as much empty space as space occupied by molecules.

A constant battle rages between the attractive forces of the water molecules, which makes them want to knit into a structured solid (ice), and the heat energy that keeps them darting, twisting, and squirming. Within the water the attractive and disruptive forces are applied equally in all directions, but at the surface of the water there is a chaotic situation. The molecules at the surface are attracted downward by the molecules below them, but have no corresponding pull from above; thus the surface layer is dragged downward, causing it to try to shrink, and creating a stretched elastic layer. The layer is so strong that a sewing needle lowered carefully onto the water surface will float on the film. Yet even while the surface is being pulled down as a whole, a speedy molecule that darts upward can pierce the layer, like a fish leaping up for food in a tank, but, unlike the fish, once the molecule leaps into the air, it has escaped the attractive forces of its neighbors in the liquid and is free to make its way with little hindrance through the sparser molecules of air.

72

Thus water loses its speedy molecules, and the slower ones are left behind as the water evaporates. Because fast molecules are hot ones, and slow molecules are cool ones, the water in the glass becomes cooler because the average speed of the molecules has been reduced. A glass of water evaporating away quietly is a beautifully balanced dynamic system. Heat flows through the glass and speeds up the water molecules, causing the faster ones to be ejected upward into the air. The remaining water is cooled, lowering its temperature below that of the room. Then, because heat always moves from higher temperatures to lower ones, more heat flows in through the glass and evaporation continues.

In nature the air above water is not confined. Water molecules evaporated from the ocean surface in one place are carried away by the wind, elsewhere they may condense into rain and fall back on the sea. To understand the real systems it helps to study the behavior of water in a closed system.

We draw a glass of water from the tap, put it under a bell-shaped jar filled with dry air, and keep any heat energy from flowing into the system. A little water evaporates into the air (a pressure gauge will show a slight increase above the initial atmospheric value) and the temperature inside the bell jar goes down a little bit. After a short time the water stops evaporating, and the temperature and pressure remain constant with the pressure of the air a little higher and the temperature a little lower than when we started. In the molecular view, fast moving water molecules have escaped from the water into the dry air, leaving slower ones behind. The addition of the water vapor to the air has increased the bombardment of the sides of the bell jar and has thus increased the pressure. When constant conditions are achieved, the water molecules escape into the air at the same rate as they are returned to the water. If we were to analyze the air we would find in it about 2½ percent water vapor at room temperature.

If we run the bell jar system at a higher temperature than before, the system will behave in the same way, but the pres-

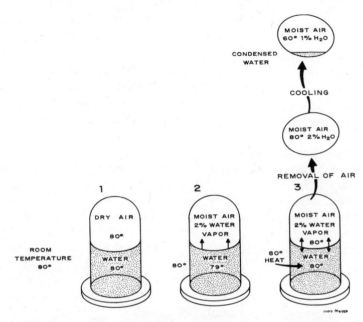

The bell jar experiment. In 1, water at room temperature is put into a bell jar with dry air above it. In 2, water molecules evaporate into the dry air until the air is saturated (at about 2% water vapor), thus cooling the remaining water. In 3, heat flows into the bell jar from the warmer room air, until the temperature is everywhere 80°. In stage 3, the air in the bell is saturated with water vapor, and its relative humidity is 100%. If the moist air is removed from the bell jar and cooled, it will eventually lose some of its water by condensation.

sure increases more before it becomes constant. The temperature goes down a little, but levels out at a higher value than before. Analysis of the air shows a higher percentage of water vapor. For each temperature there is a definite water vapor content of the air, which increases rapidly with rising temperature. This is the content in equilibrium with the water. When air achieves equilibrium content it is said to be saturated, and the relative humidity is 100 percent. If only half the equilibrium amount is present the relative humidity is 50 percent.

The amount of water vapor the air can hold increases with increasing temperature, so if we saturate air with water vapor, then remove heat and lower the temperature, the capacity of the air to hold water vapor is diminished and the excess condenses back into liquid water.

Finally, if we put a piece of ice under the bell jar full of dry air, at a temperature far below freezing, we find that solid ice has a vapor pressure just as water does. Water vapor from the ice enters the dry air and increases to a fixed amount at a fixed temperature. As the temperature of the bell jar is increased, the amount of water vapor in the air increases, and the amount for a given temperature is always the same. So it is possible to talk about relative humidity with respect to just air and ice. If the amount of water vapor in subfreezing air is less than the fixed amount that would be present at that temperature in a closed system of ice and air, the relative humidity is less than 100 percent. The amount of water vapor that can be held by subzero air is small compared to the amount that can be held by room temperature air—only a fraction of one percent, as opposed to the 2½ percent it can hold at 70° F.

When a system saturated with water vapor is cooled and water condenses out, much heat is released. If the procedure is reversed and water evaporates into dry, warm air, an equal amount of heat has to be put into the system. This large energy addition or subtraction is the basic reason for our current lack of hope of major weather modification. The energy required to change natural systems overwhelms even the potentials of nuclear power. A single room twenty feet long, fifteen feet wide, and nine feet high contains about two hundred pounds of air. On a humid summer day the air would hold about six pounds of water vapor, or, if it were condensed, about three quarts of water. Even at 100 percent efficiency, airconditioning equipment powered by a one horsepower electric motor would have to operate for three hours to condense all of the six pounds of water vapor from the air. This example gives some idea of the amount of power and equipment that would be necessary to make a substantial humidity change in an outdoor system, even for a small city park.

Many of our everyday experiences are dominated by the evaporation and condensation of water, with the large related heat change. In our own personal heating and cooling systems, we find that our body temperature of ninety eight and six tenths degrees is a most convenient one. Body temperature is higher than air temperature almost all the time. When seventy degree air, even if it is saturated with water vapor, blows against the skin, it is not saturated with water vapor at skin temperature. Water then evaporates from our skin into the air, taking up heat energy and cooling us. True enough, every once in a while we run into a situation in which we are exposed to air saturated with water vapor at our body temperature and then we really suffer. The Gulf Coast of the United States, during the summer, is famed for the degree of discomfort created by its weather, even though the air temperatures are not as high as in many other places. The air often comes from the Gulf of Mexico where it has idled its way and become close to saturation at the temperature of the warm Gulf waters. When it arrives at the home of a perspiring resident of Houston, Texas, it can take a little more water vapor from him, but not much. Without air-conditioning, he finds that his personal cooling mechanism is not sufficient for his comfort.

In the deserts, air temperatures can reach one hundred and forty degrees, but this may not produce as much discomfort as ninety degrees on the Gulf Coast. The air of the deserts has been heated many degrees in the absence of water, so it is avid for moisture. Its relative humidity may be almost unmeasurably low. Sometimes desert air has less than 1 percent of the water vapor it would take on if given a satisfactory source. When it encounters a person, any moisture on his skin is evaporated rapidly, with excellent cooling effects. The danger, of course, is excessive water loss.

An interesting phenomenon occurs in tropical countries with high temperatures and high relative humidity. Iced drinks are understandably popular, but they sometimes are a nuisance, for the ice seems to melt almost instantaneously, and the glass sweats copiously, preferably onto someone's fine mahogany table. What happens is that the air is nearly saturated with water vapor at a high temperature; when it comes in contact with the cold glass it is cooled far below the temperature of saturation. Water condenses on the outside of the glass and releases a great deal of heat energy which is then available to melt the ice inside. A beverage-filled glass in the tropics is an efficient little mechanism for melting any ice that is added.

Many more interesting phenomena have similar explanations. A dishtowel hanging outside in subzero weather will "freeze dry" as the frozen water evaporates directly to vapor, even though it is as stiff as a board during the whole process. The wind blowing on the dishtowel must not be saturated with respect to ice and water vapor.

When water vapor in warm air blows against a cold window pane the water molecules in the air are slowed down and form a crystalline coating of ice on the glass. One of our friends learned about the changes in the physical states of water by a harsh experience. He put an air controlling system into his house, but neglected to insulate his attic. The system seemed to work well; his house was well ventilated with warm moist air throughout a long cold winter. Finally spring came, and on the first warm day he noticed water dripping down through all his second floor ceilings. He ran up to the attic, where to his horror he found what looked like an ice cave—several tons of frost had accumulated as the warm moist air from below had condensed directly to frost and had painted a gigantic mural beneath his roof. All went well until the frost began to melt, and in the end he had to stand by and watch his plastered ceilings fall one by one.

Our friend was trying to avoid the dry indoor air of winter time, especially when the outside temperature is very low. Even if the air is saturated with water vapor at twenty degrees below zero (in equilibrium with snow), it contains only about 5 percent as much water vapor as it can hold at a room temperature of seventy-five degrees. Without some kind of artificial humidification, the dry air picks up moisture wherever it can. People are prime targets. Moreover, because the air is so dry and evaporates water so readily, skin temperatures are lowered below the air temperature. We feel chilly and turn up the furnace, making the air drier and making us feel even colder. This kind of relation is the basis for the development of several kinds of "comfort" or "discomfort" indices, based on air temperature, relative humidity, wind velocity, and so on. Air temperature alone is not a very good criterion of our personal heat loss, which must be kept within limits if we are to be comfortable.

In deserts, despite modern equipment, one still sees the old fashioned air cooler made from a piece of burlap dipped partially into a tub of water, with a fan blowing air from outdoors through the burlap. The burlap acts as a wick; it stays wet as water creeps up it by capillary action. The dry, hot air, passing through the wet burlap, evaporates water and gives up heat to change liquid water into water vapor. The air is cooled and its moisture content is increased. Because the air is so hot and dry to begin with, it is still far from saturated with water vapor at its new lower temperature. It can still evaporate perspiration from the skin and keep the body comfortable.

An interesting discovery about the behavior of water vapor is that it will not condense into water droplets unless there are dust particles or violent disturbances to give the droplets a start. Warm, clean air can be put in contact with liquid water so that it becomes saturated with water vapor, then the air can be removed and cooled. With impurities in the air, a cloud of water drops is formed; with clean air nothing happens, and it is possible to lower the air temperature far below freezing. If dust is then introduced, the dust particles become covered with ice. In the high cold parts of the atmosphere there is commonly a lot of water vapor waiting to turn into ice. When a jet plane flys along in temperatures of thirty to forty degrees below zero, the disturbance caused by the jet performs the same role as dust particles. The familiar jet trails are streamers of ice crystals. It has been said that jet trails are now a significant fraction of the clouds over the United States.

A peculiar property of the water droplets in the white clouds sailing overhead is their unwillingness to join together to make raindrops. A big drop is ordinarily preferred by nature to a small drop. It would seem logical that every time two drops jostle each other they should unite to form a large one. Instead they refuse to do so, perhaps because each droplet has a slight electrical charge at its surface which causes it to repel another approaching droplet. Just what controls the size of drops is still quite a mystery.

Every white cottony cloud is a potential source of rain, provided the droplets can be encouraged to join together. A great deal of experimental work has shown that crystals of silver iodide or of dry ice are effective in helping the droplets to overcome the repulsion of their electrical charges. Thus cloud seeding was born. Making rain by seeding clouds with finely ground silver iodide scattered from an airplane has been clearly successful in many cases and a failure in many others. Control of rain by seeding has suffered more from legal than from financial difficulties however. It is the old problem of water rights—who owns the water in a river that flows from one city or state to the next? Who owns the rain stored in the passing clouds? The legal aspects of cloud seeding are even more complex than those of river use. Who can prove that a cloud might not just have disappeared by itself and never given anyone a shower? Or who can prove that the change in the amount of sunshine that someone gets, because protecting clouds would have formed if someone hadn't seeded clouds somewhere else, is right or wrong? Cloud seeders have been accused of causing deluges and not just beneficial rains, of making it rain in the wrong places, of claiming to have made rain when it would have rained anyway.

We can predict that seeding eventually will be put on a firm and controlled basis, but it is going to take basic scientific work and further development of techniques, plus political and social change, before rainfall can be systematically controlled. There are also important psychological factors: just as many people feel upset at having the moon tampered with, they also object to harnessing clouds, which they feel have some inalienable right to form and float and disappear in response to natural forces. The jet trails give us the warning of things to come. We wonder if anyone has the same disturbing feelings about turning a jet trail into rain as he does about doing the same thing to a naturally formed cloud.

The kinds of clouds we see fall easily into our general picture of water, ice, and water vapor. On many summer days we see flat-bottomed clouds, with their lower edges all at the same level, and often a little darkish, with a billowy cream puff top. On such days the lower air is warm and moist. The sun heats the ground which in turn heats the air; the air warms, expands, and rises. As it rises it cools and at some height is cool enough to cause the water vapor, encouraged by dust particles as nuclei, to condense into droplets, forming the flat lower surface of the clouds. With condensation there is heat release, and air is further heated and expanded. The rate of rise accelerates, strong upward air currents are generated, colder air above is displaced and moves down. The interior of the cloud becomes a highly active zone; rising and sinking air mix in great swirls and eddies. Any airline passenger who has flown into the white billowing heart of one of these clouds knows the motions the plane performs in response to turbulent air.

There are few sights more awesome than flying between these great cumulus clouds when the conditions that cause them are extreme. The interior of the cloud is a seething mass of grey and white, lightning flickers ominously as coalescence of the droplets is accelerated by the fierce internal winds. Almost incessant electrical discharge takes place. The flashes are diffused by the intervening cloud so that the observer sees eerie reddish glows. The fast-rising air climbs many miles and is quickly chilled far below zero; ice crystals form, and may grow as they

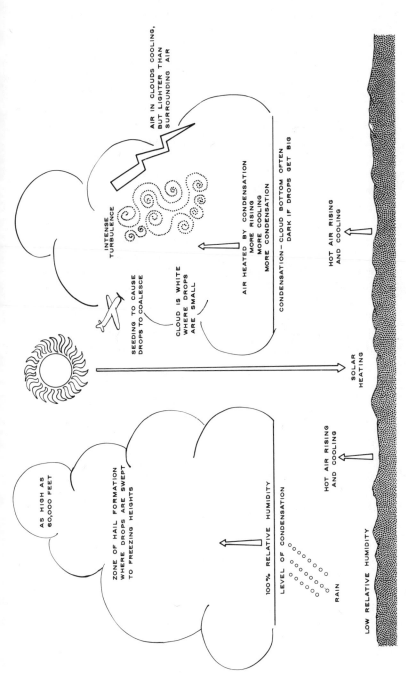

Formation of clouds

AIR IN CLOUDS COOLING, BUT LIGHTER THAN SURROUNDING AIR

INTENSE TURBULENCE

SEEDING TO CAUSE DROPS TO COALESCE

CLOUD IS WHITE WHERE DROPS ARE SMALL

AIR HEATED BY CONDENSATION
MORE RISING
MORE COOLING
MORE CONDENSATION

CONDENSATION—CLOUD BOTTOM OFTEN DARK IF DROPS GET BIG

HOT AIR RISING AND COOLING

SOLAR HEATING

AS HIGH AS 60,000 FEET

ZONE OF HAIL FORMATION WHERE DROPS ARE SWEPT TO FREEZING HEIGHTS

100% RELATIVE HUMIDITY

LEVEL OF CONDENSATION

RAIN

HOT AIR RISING AND COOLING

LOW RELATIVE HUMIDITY

fall, then are swept up to fall again and possibly to grow into huge hailstones. Some parts of the cloud are serene; there are tunnels and caves of clear air, elsewhere angry grey wisps of cloud circle and twist. At the very top of the cloud it is sublimely white, expanding like a multiple mushroom, beautiful to see from below, deadly to enter by plane.

Under such conditions tornadoes are born. We still do not know quite why. From the grey seething underside of the turbulent clouds a finger of cloud reaches down, a hollow, twisted, spinning finger. The tip of the finger sometimes touches the ground, sometimes it lifts above, but wherever it touches it leaves a devastation that is almost incomprehensible. The tornado moves across the landscape slowly, ten to twenty miles an hour, its probing finger like some huge fantastic creature reaching down from above the skies to probe the surface of the earth. But if the aimless wandering of the finger is slow, its spin is not. The vortex is but a few hundred feet or yards across. The winds chasing around this little circle may blow more than two hundred miles an hour.

Tornadoes take many shapes as the spinning funnel moves along. If the funnel is two hundred feet across and the winds are two hundred miles an hour, the funnel makes a full turn in about three quarters of a second. Inside the funnel pressure is low, outside it is normal. The tornado acts like a giant vacuum cleaner, sucking everything in its path up into the center of the vortex.

Tornadoes have picked people up, carried them a quarter of a mile, and deposited them unhurt. They have lifted automobiles and dropped them through the roofs of houses. They have driven sticks through telephone poles. Their energy comes from the intensification of a simple, everyday process. The air becomes less dense and rises through the cooler, heavier air above. As it rises, it expands and moves into lower pressures, and thus rises still faster. So we have a natural system, which, once started, continues to accelerate in its motions.

Hurricanes are like tornadoes, except that they form over the warm waters of tropical seas and cover an area of many hundreds of square miles. Just what causes the ordinary convection of the air warmed and humidified at the sea surface to become organized into a gigantic whirling storm is not clear. But once organized it maintains and propagates itself by the same process as the tornado. Once a chimney-like column of warm air starts rising, and more air moves in across the sea to become warmed, humidified, and lightened, the storm builds in intensity. The faster the air rises, the more air rushes in to take its place, until strong winds, commonly of a hundred and twenty-five to a hundred and fifty miles an hour, are generated. Hurricanes are finally defeated either by moving across land, where their force is gradually spent, or by moving into areas of colder waters which cool the air and slow the strong upward convection that drives the storm center.

The amount of precipitation accompanying a hurricane is a major cause of damage on land. The cooling of the great mass of moist air, especially as the storm dies out, may dump ten inches or more of rain in one spot in a few hours. Hurricane damage in the United States has been increasing by leaps and bounds since the turn of the century, not because the force or frequency of storms have increased, but chiefly because there are more buildings and denser populations in hurricane pathways. Until 1950, the annual damage was estimated at about one hundred million dollars; since then it has roughly tripled. From 1965 through 1969, the average yearly damage was about five hundred million dollars. Thanks to better warning systems and more information about the nature of the storms and accompanying high tides, the number of deaths has dropped as abruptly as dollar damage has increased, but the toll is still greater than necessary. Before 1950 an average of three hundred people were killed a year in hurricanes; since that time the number has dropped to less than one hundred yearly.

Cloud-seeding experiments have been tried in an attempt to reduce hurricane intensity. Hurricane Debbie was seeded in 1969 as a project of the National Hurricane Research Laboratory. The wind speed is related to the large temperature difference between the warm air of the hurricane center and that outside. By seeding the air on the edge of the warm mass and causing precipitation of supercooled air, it was hoped that enough heat would be released to lessen the temperature differences and thus dissipate the storm. Although it is difficult, because of the large rapid natural variation in intensity of hurricanes, to be sure that the seeding was responsible, a diminution of about 15 percent in the maximum wind speeds resulted, apparently from the seeding experiment.

The temperature of the lower atmosphere diminishes about 3° F for every thousand vertical feet. At about seven miles it becomes constant at thirty to forty degrees below zero F, and remains at that temperature for another five vertical miles. Above twelve or thirteen miles the temperature begins to go up again, and from there up to thirty miles increases back up to freezing. Then it goes steadily down again, dropping to about fifty degrees below zero at fifty miles up.

A fairly common condition in the lower atmosphere is to have clouds made of water droplets that have been chilled well below their freezing temperature, but have stubbornly resisted the change to snow or ice. When ice does form in the atmosphere, it somehow manages to crystallize into the beautiful hexagonal crystals of snow. There are few more beautiful sights than a steady light snowfall of big flakes, some of them thin platelets as much as half an inch across. It still seems to be true that no two snowflakes are identical. Snow crystals are so feathery and pack so loosely that it takes nearly ten inches of snow to equal an inch of rain.

Whether the glacial age is over or not is an interesting question. If the merest trace of snow manages to remain through each summer, accumulation begins. In the Rocky Mountains of Colorado and Wyoming, snow on the high peaks usually disappears late in June, and the next snowfall that stays comes in late September. But in years of heavy winter snowfall and cool summers, a fair amount of snow may remain until late July or early August. It wouldn't take much change in the amount of winter snow, or in summer temperatures, to begin building glaciers again in the high mountains—large ones compared to the few small ones that still remain. Like so many of the natural processes that have been described, accumulation of snow is a selfperpetuating phenomenon. Once started, like a hurricane or a tornado, it creates new conditions favorable for its survival. In years of a heavy snow blanket, in the northern United States for example, the presence of the snow causes reflection of sunlight which slows the spring thaw. The snow also uses heat in melting, without permitting temperature change to above freezing, and provides water vapor to the air, all of which processes produce conditions favorable for more snow and the retention of what is already there.

When snow does accumulate and thicken, it soon changes into granular ice and finally into massive ice that becomes a glacier. The snow line, that height above which snow accumulates and glaciers form, is reached by high mountains even at the equator. There are glaciers on the equator in the Andes mountains of western South America. In the Rocky Mountains of the United States the snow line is at about fourteen thousand feet, close to the tops of the highest peaks; it gradually descends until at the North Pole it is at sea level.

The ice caps of Antarctica and of Greenland today are the only important remnants of the Great Ice Age, with the exception of numerous but relatively insignificant mountain glaciers in the high latitudes. The entire ice mass, if melted, would raise the oceans two hundred feet, which corresponds to an ice mass about 1½ percent of the mass of the oceans.

The Greenland glacier has accumulated on a mountainous surface and has buried all but the tops of the highest peaks, which stick up here and there through the ice as nubbins of rock. The ice is a mile thick in many places and caps Greenland with a gently curving white dome. The ice behaves like a huge pancake poured onto a griddle; the glacier is fed by snow on top and discharges by flowing and spreading at the edges. At the margins of Greenland, and especially on the western edge, the ice flow is channeled down valleys that it has scoured and widened into great troughs by centuries of growth and movement. When the ice comes to the sea, great chunks break off to make icebergs that are buoyed up by the seawater. The sea surface to the south and west is dotted by hundreds of huge bergs. As many as fifteen thousand are produced each year. Many icebergs become stranded near their source and melt. In summer the water from melting ice is enough to lower the salinity of the seawater of the North Atlantic, and the influence lasts throughout the year. Other icebergs work their way irregularly north and then move southward in the Labrador current; some 400 yearly menace North Atlantic shipping. Their paths are now plotted by the International Ice Patrol, a warning system of ships, planes, and radar, which has proved most successful. Small ships are still wrecked from time to time by colliding with icebergs in the fog, but no disaster comparable to that of the *Titanic*, which, on her maiden voyage and after wide publicity about her unsinkable design, struck an iceberg and went down with 1,517 persons aboard, has occurred.

When the global relations of evaporation, condensation, and precipitation are surveyed, a clear-cut general picture can be drawn. It is useless to try to interfere, even with the most advanced technology, even with the use of atomic power, with the basic energy relations of water and the atmosphere. Evaporation of water from the sea, followed by condensation and precipitation of the water vapor, is a natural cycle that involves total energy use beyond our abilities to modify. The energy received from the sun is enormous in its total, but sufficiently diffuse in its local effects to prevent its efficient collection and employment for widespread use as power. Nor can a way be anticipated of collecting it and using it to modify importantly its own global effects.

The great hope in managing precipitation comes from research on ways to control the natural pathways of energy release. Fog could never occur if the water droplets would co-

alesce and fall to the ground as rain. This is what they should
do, and would do, in a world with the patience to wait forever.
As has been said, the usual white clouds overhead would not
be there if their droplets could find a way to unite and release
their energy. When water is evaporated by the sun, then rises
as water vapor, reaches saturation, and condenses back into
liquid water, it produces a wide variety of conditions, almost
all of which are transitory. Given time, condensed water will
return to the ground. But what happens between initial condensation and return to the earth as liquid water, that can filter
into the soil and collect into streams, is an infinite variety of
intermediate states, in which the individual droplets refuse to
get big enough to fall. A white cloud is like a bomb waiting
to explode, in the sense that it has large amounts of energy
waiting to be released.

The future of weather modification rests on finding the right
catalysts to help natural processes occur faster. Cloud seeding
by silver iodide is just one method, out of a thousand possible
but untested ones, that has been made to work. It gives us a
peek into the future, in which it may be possible to dispense
with or manage, clouds, fogs, tornadoes, and hurricanes. All of
these can be controlled if it can be shown how to help small
drops become larger ones. This is the way Nature wants to go;
all that has to be done is to find out how to accelerate the
natural process. The enigma remains: can it be done? Experience says yes, if enough painstaking work is supported. But the
support must be the support of hope, and cannot be guaranteed
to be successful.

~~~~~~

CHAPTER FIVE

Rivers

THE RIVERS of the world discharge 35 trillion tons of water to the oceans every year—about ten thousand tons for every man, woman, and child now living. Presumably this share will diminish to five thousand tons per person by the year 2000 when the population will have doubled. Five thousand tons per year is many times the present average use for the world, but only about twice the amount used per person in the United States. The day may come, a few generations hence, when all of the water of all the rivers will be withdrawn at least once during their course to the sea.

Future river management and utilization will focus on twenty major rivers that deliver over 50 percent of the water added to the oceans. The first ten rivers of the world, ranked by their discharge, are the Amazon, the Congo, the Yangtze, the Brahmaputra, the Ganges, the Yenisei, the Mississippi, the Orinoco, the Lena, and the Parana. The exact order is suspect because of the uncertainties in the information about discharge. If they are ranked on drainage area, the order changes considerably. The Mississippi then jumps from seventh to third and the Yenisei from sixth to fourth.

The Amazon overshadows all the others with an estimated 7 million cubic feet per second (cfs), more than ten times the Mississippi discharge of 630,000 cfs and about a fifth of the total for all streams! The Congo is next, with a million four hundred thousand cubic feet per second. Of the next eight, the Yenisei and the Lena in the USSR, the Orinoco and the Parana in South America, the Brahmaputra in East Pakistan, the Ganges in India, the Yangtze in China, and the Mississippi all discharge about the same amount—between six and seven hundred thousand cubic feet per second.

The Amazon and the Congo owe their preeminence not only to huge drainage basins, but also to their locations close to the equator in zones of heavy rainfall. Only recently have three great northern flowing rivers of the USSR, the Yenisei, the Ob, and the Pechora, received much publicity. These streams drain the vast lowlands of Siberia, and it is almost a happenstance that they drain to the Arctic, for the slope of the land to the north is almost imperceptibly greater than that to the south. The Rusians have suggested the possibility of reversing their flow, so that their waters, now mostly wasted on the frozen north, could be used by the highly populated areas to the south. Needless to say, the consequences of diverting the Ob, for example, into the Black Sea, which connects with the Mediterranean, would change the nature of both the Black Sea and the Mediterranean, and would affect the lives of people in many different countries, for the present flow of the Ob is only a little less than that of the Mississippi. According to Hubert Lamb, a scientist of Britain's Meteorological Office, about half of the fresh water now flowing to the Arctic would be diverted by the Russian plan. About eighty-five million acres could be irrigated and a hundred and fifty million acres of swampy land could be drained, a total area about 10 percent of the size of the United States. Apparently the Russians would divert the rivers into the

Caspian and Aral Seas, rather than into the Black Sea. No matter where they are diverted to the south, the effects on world climate will be drastic, though hard to predict specifically. In general, the effect would be to shift warm climates northward, so that France might become more like Morocco is today. Also, the fresh water now flowing to the Arctic dilutes the seawater there and makes it easier to freeze. Lacking the flow from the Russian rivers, part of the Arctic ice pack might melt. Inability to predict the consequences in any detail has heretofore prevented the Russians from putting the plan into action, although recent reports suggest that work may have started.

The great rivers of southeast Asia, the Ganges, the Brahmaputra, and the Yangtze, head in high mountains. They carry vast amounts of mud and sand. Strong efforts are being made to manage them more successfully, to control flooding and silting, and to make better use of their flow.

Oddly enough, the water that runs off each square mile of land averages about the same for every continent. North America and Asia are estimated to contribute the same amount per square mile and to be about "average" on a global basis. South America is somewhat wetter than the average continent, thanks to its tremendous area of tropical climate. Africa feels the influence of its great deserts and thus is below the world average, despite the high runoff from the tropical Congo basin. The Nile has a drainage area almost as large as that of the Mississippi but discharges only one fifth as much water to the Mediterranean as the Mississippi does to the Gulf of Mexico. Europe is like Asia and North America. The continents are vast enough, and extend far enough north and south so that local differences in climate tend to be evened out. The major rivers carry the water of many tributaries, some of which drain deserts; others come from well-watered lands.

The continents are being slowly eaten away by water. The amounts of material carried dissolved in streams, or swept along as suspended sand or mud, do not sound impressive when compared to the volume of water. The average river carries only about 140 parts per million dissolved material and perhaps 500 ppm solid particles. Each gallon of water carries only one fiftieth of an ounce of dissolved matter and four fiftieths of an ounce of sand and mud, but these add up to around twenty-three billion tons of continental materials carried to the seas each year—more than four hundred tons for each square mile of land surface. This overall average includes Australia, which is so dry that it is hardly being denuded at all, as well as frozen Antarctica.

The United States, with its three million square miles, is typical, annually losing an average of eighty-five tons per square mile as dissolved material, and two hundred and fifty tons of sand and mud. The whole country is being lowered by about one foot each ten thousand years.

South America, despite its high rainfall and large water discharge per square mile, is not being lowered quite as fast as the other continents. Its waters carry a light load of dissolved and suspended matter. Europe is the most soluble continent; one hundred and twenty tons per square mile are being dissolved away yearly. But Europe may be a good place to live in the long run; because the sand and mud carried to the ocean add up to less than a hundred tons per square mile, it is being lowered at two thirds the rate of the United States.

The area that is disappearing fastest into the ocean is Southeast Asia. Eighty percent of all the sand and mud carried into the world oceans comes from that area. The rate of "mechanical" denudation is more than a thousand tons per square mile per year. The combination of high rainfall, high mountains, and extensive cultivation all aid in the washing away of the soils. To make it worse, in China there are extensive areas underlain by thick layers of wind-blown dust. When cultivated, they are highly susceptible to erosion. The Yellow River gets its name from its high content of this dust.

Although the total amount of rain on the continents has probably not changed very much for a long time, erosion has been increased more and more by the activities of man. An estimate has been made by Sheldon Judson at Princeton University that the rate of lowering of the continents has been doubled by human influences. A change from grassland or forest land to crop land increases the rate of erosion many times. During construction, a site for a new home or for a freeway may erode a thousand times faster than the original wooded land.

The United States has sufficient area and enough range of climate to serve as a model of river drainage conditions and problems throughout the world. Its only serious deficiency as a general model is lack of a major tropical drainage basin like that of the Amazon or the Congo. The total area of the United States, excluding Hawaii and Alaska, is three million square miles, with a total runoff of about one million eight hundred thousand cubic feet of water per second. This corresponds to nine inches of rain annually, or a third of the average annual precipitation—just about the world average. The other two thirds is lost by evaporation. There are two hundred thousand square miles of desert country in the Southwest with interior drainage. They contribute no water to the sea.

Water supply problems in the United States are reflected in the river flows of the various regions. The major rivers of the Atlantic slope, from New England to Florida, discharge about 350,000 cubic feet per second; the Mississippi, draining the central interior, sends over 600,000 cubic feet per second into the Gulf of Mexico. The Gulf is indeed the favored recipient of the United States stream water; the coastal rivers east of the Mississippi, draining the humid southeastern states, add almost 200,000 cubic feet more. West of the Mississippi evaporation is high and rainfall is only about half as great as in the southeastern states, with the result that a drainage area of four hundred thousand square miles contributes only fifty-six thousand cubic feet of water per second to the Gulf. The Rio Grande, famous as it is, and draining nearly two hundred thousand square miles, discharges only three thousand cubic feet per second.

The Colorado River, with its quarter of a million square miles of drainage area, has a discharge of a little less than 1 percent of that of the Mississippi. Yet its importance in southwestern

United States should not be underestimated; in the land of the desert, even a small river is king. No wonder there are water problems, when a third of the whole United States—an area of a million square miles—from the Sierras on the west, the Mexican border on the south, and eastern Texas to northern Nevada, has a total discharge of sixty thousand cubic feet per second. The area includes most of eight large states and is roughly the same size as the Mississippi drainage basin.

West of the Sierras, the Pacific slope, like the Atlantic slope, is drained by many medium and small-sized rivers. Because of the high precipitation in the mountains, the area as a whole can be classified as well-watered. Discharge is about 20 percent greater per square mile of drainage area than for the Mississippi basin. The northwest, drained mostly by the Columbia River, is in good shape. Discharge is about two thirds that of the Mississippi from an area only one fourth as large.

This brief summary of stream discharge highlights the water shortage in the southwest. The shortage problem is aggravated by the poor quality of the water. Naturally high evaporation, increased by extensive irrigation, raises the salt content of the water. The Colorado River, source of most of the available stream water, carries 750 ppm or more dissolved salts in its lower course. In contrast, waters of the western Gulf of Mexico drainage average about 250 ppm salts. The Mississippi, at 200 ppm, is well above the world average of 140. The softest waters flow down the rivers of North and South Carolina and Georgia. With only 55 ppm salts, they have been most attractive to many industries and a pleasure to individual users who like a good lather in the bath or shower. The eastern Gulf states and the Columbia drainage area also have water of good quality, with about 90 ppm dissolved solids.

The chemistry of the major rivers of the world is remarkably similar. Just as their large drainage areas tend to make their flows the same for each square mile drained, their chemistry also is a composite of all the different kinds of rocks and soils sampled by their many tributaries. The elements they acquire are mixed into the master stream. River waters are dominated by calcium and bicarbonate, in contrast to domination of ocean waters by sodium and chloride. Calcium and bicarbonate are traceable to the limestone bedrocks in the source areas, and show the cyclical nature of the movement of earth materials. Most of the limestones were originally deposited in the sea, their calcium supplied by forgotten rivers of the past that got *their* calcium from limestone deposited in still more ancient seas. There may have been as many as ten such turnovers of calcium throughout the immensity of geologic time. No wonder the early geologist, James Hutton, cried out that there was no beginning and no end to the transformation of sea bottom into land and land into sea.

Bicarbonate makes up almost half the dissolved load of streams, averaging 60 parts per million. Calcium is 15 ppm, silica 13, sulfate 11, sodium and magnesium and chloride 5–10, and potassium 1 or 2. These substances make up 99 percent of the dissolved elements in river water, just as they make up 99 percent of dissolved elements in sea water, but their concentrations are far less and their ratios to each other are quite different.

It can almost be said that hard rocks make soft waters, and vice versa. The North Atlantic slope owes its soft dilute river waters to a combination of high rainfall, low evaporation, and bedrock that resists the solvent action of the rain. There are few limestones or deposits of gypsum in the area to yield calcium, magnesium, and sulfate to groundwaters percolating into the streams. The rocks are very old and have been tempered by heat and pressure so that few easily-won constituents are left to be carried away by the rain.

The South Atlantic slope has even softer water. The warmer climate and the gentle topography have permitted leaching of the soils so that now, after a rain, when the dilute overland flow has ceased, the water trickling into the streams from the soils finds little to dissolve.

The waters of the Mississippi are truly cosmopolitan. The

Missouri, with its headwaters far to the north and west in the Rockies, flows through the soft rocks of the semiarid states of the Great Plains and has a fairly high salinity of 260 ppm. Most of the rocks are geologically young and contain much chloride and sulfate that can be easily leached away. Evaporation in the Great Plains is high; the rain waters of the summer storms and the meltwaters of the snows from winter blizzards sink into the soil and are evaporated out again, bringing soluble materials up to the surface where they can be washed into the streams by the great spring storms. The Muddy Mo, or the Big Red as it is sometimes called, has an extremely irregular discharge. At times of high water it carries a great burden of suspended mud washed from the soft rocks of its drainage area. The Platte River, draining finally into the Missouri at Omaha from the even more arid region south of the Missouri drainage, has the characteristics of the Missouri, though more pronounced. Its discharge fluctuates wildly with the cloudbursts of summer. The Platte and its North and South Forks move large quantities of sand and mud during storms. When the storms subside, sand and mud choke the channels. As there is little water stored in the soil to feed the stream, what is left wanders in a net of rivulets among the flood debris.

To the east, the Ohio, Tennessee, and Cumberland Rivers flow into the Mississippi from the well-watered western slopes of the Appalachian mountains. A hundred and fifty years ago their waters were probably soft and dilute. Today they are strongly influenced by industry and mining and sheer population pressure. In the Ohio drainage area especially, the mining of coal has created a serious and so far insoluble problem. The wastes from coal mines are high in pyrite, a mineral containing iron and sulfur. While pyrite is safely entombed in the rocks away from oxygen it is perfectly happy to remain as iron sulfide, but when air is admitted by mining, iron oxide and sulfuric acid are produced. The piles of mining waste, and the abandoned mines themselves, continue to produce sulfuric acid which seeps into the streams and makes them acid, as well as increases their sulfate content. There are more than sixty thousand individual sources of polluted waters from mining operations in the Ohio drainage basin.

The sulfuric acid kills fish and changes the whole environment for aquatic plants; the seepage from mines also consumes oxygen from the stream waters into which it flows. The problem is a difficult one. Most attempts to seal old mines have been unsuccessful; once the ground is opened up it seems that resealing is almost impossible. In some waste piles the reaction of oxygen from the air with the iron sulfide in the waste generates enough heat to cause the waste to burst into flame, releasing various noxious sulfur gases to the atmosphere. Much research has been done in trying to stop the acid sulfate contamination in abandoned or in active coal mining areas, but so far there has been no great technologic breakthrough.

The extensive coal and steel complex in the Ohio drainage area, including Pittsburgh and Youngstown, has not helped the situation. The effluents from steel plants, such as strongly acid pickling liquors, have had horrid effects on the rivers. Yet there is hope. The city of Pittsburgh is working hard to minimize direct industrial contamination of streams and of the atmosphere. Much of the pollution is being stopped by ending the direct dumping of industrial wastes into the waters.

The rivers of the southwest, like the Rio Grande and the Colorado, drain areas of sparse vegetation where rain tends to come in sudden, violent storms. These conditions are ideal for creating salty, muddy water. During floods, the Rio Grande carries a thick brownish soup. Members of one of the early expeditions down the Canyon of the Colorado found that their western sombreros had an unexpected use—they could be used as settling basins to make the muddy water fit to drink. The rivers clear up during low flow when the water is coming from the soil, but they remain salty because the soil waters have collected salt during the periods of evaporation between rains. The importance of the Colorado and the Rio Grande should not be minimized, even though their quality today is poor and the kinds and quantities of crops that can be irrigated with them is now restricted. More irrigation will make them saltier still. Chemically they are high in the common elements, and their levels of magnesium and sulfate are high enough to cause digestive upsets.

The Columbia River is the great reservoir of unused fresh water in the United States and is being eyed greedily by Californians. The eighty thousand cubic feet per second they have in their own streams is not enough for the growing needs. Eyes still turn north toward the Columbia. Its drainage area is well watered, the rocks are hard and not easily dissolved, the water is fresh and cool, and the supply—at the moment, over four hundred thousand cubic feet per second—is discharged largely untouched into the Pacific Ocean.

This tour of the United States illustrates the basic principle that areas of high rainfall and low evaporation and with rocks resistant to chemical solution, like those of most of the Atlantic slope, have abundant, good quality water. In arid areas with low rainfall, high evaporation, and especially if they have soft, geologically-young rocks, the salt content of the rivers is high, sediment load is heavy, and discharge tends to be important only after violent storms in the drainage area. The transition zone between abundant or adequate good water and poor quality water in inadequate amounts is rather sharp. As discharge drops, quality worsens. If the reverse were true, many problems would be ameliorated.

The sequence of events that occurs during and after a rain storm in the drainage area of a stream has been worked out in detail in a fine piece of work by V. C. Kennedy of the Water Resources Division of the U.S. Geological Survey.* Kennedy studied the Mattole River basin in northern California that shows behavior typical of many streams. At low water the stream carries little suspended sand or mud and the dissolved

* V. C. Kennedy, "Silica Variation in Stream Waters with Time and Discharge," in Non-equilibrium Systems and Processes in Natural Water Chemistry, *Advances in Chemistry*, 106 (1971), 94–130.

chemical content is about average for many rivers of the world. When a rainstorm comes, the stream begins to rise and in a heavy storm discharge increases as much as 500 times over low water flow. As the stream rises and speeds up, its load of suspended sand and mud increases as well, rising from almost nothing to several thousand parts per million—ten times the world average.

This means that one big storm can erode more in a few days than is eroded during all the rest of the year, and shows why it is so difficult to get accurate measurements of sediment load in streams.

Kennedy found that as discharge and suspended load increased, the concentration of dissolved salts decreased, dropping to as little as one fourth of the original amount. Even so, the total dissolved material carried is many times greater at high discharge than at low discharge, because of the terrific increase in water volume. High discharge is reached within a day of the onset of heaviest rain in the Mattole basin; in larger river systems several days may elapse before high water flow reflects heavy rain. On the Mattole, after the rain stops, discharge diminishes slowly over a period of a week or two, toward a typical low level, while the concentration of dissolved salts rises back toward the low water amount.

These changes are accounted for as follows. At the start of a heavy storm rain first soaks into the soil, then as rain falls faster than the soil can absorb it, the excess water washes over the ground surface, eroding the soil of the valley slopes. As not much rain has time to sink into the ground, the water working into the stream is chiefly rain rather than groundwater, and is loaded with mud. Because rain is so pure, the concentration of dissolved matter in the streams drops. As the rain continues, some that had sunk into the soil begins to percolate out again downslope, but in its journey through the shallow soil it picks up dissolved material. The water in the stream is now a mixture of "overland" water and water that has passed through shallow soil. After the rain stops, water continues to move out of the soil into the stream. The soils of the hillsides behave like a bathtub with a small leak—despite leakage the pores of the soil fill up during heavy rain and gradually empty afterwards. It is this "bathtub" effect that keeps streams flowing between rains. The longer the time between rains, the longer the water that feeds the river has been in the soil, and the deeper it has penetrated the soil, so that the stream water at low flow is highest in dissolved salts.

Kennedy's work documents in detail general concepts that have been proposed before and gives us much new information highly pertinent to problems of pollution and waste disposal. Because dissolved sulfate in the stream increased simultaneously with discharge and suspended sediment during one storm, he was led to suspect that the stream was receiving a lot of sulfate that had settled out of the atmosphere onto the land surface. Some other elements from the atmosphere, like chloride, are moved down into the soil during light rains that soak into the soil and cause little runoff. They appear in the streams after the maximum discharge.

Still other soil elements, released by weathering deep in the soil, get into the streams chiefly at low flow when the deep soil waters are tapped. Studies like Kennedy's permit prediction of how long it will take contaminants in soil waters to work their way into streams.

It seems inevitable that the river systems of the United States will soon have to be managed almost entirely; but most attempts to manage major river systems on a long-term basis work in opposition to important natural forces that are inherent in the nature of the drainage of water from high places to low ones. Major streams tend to silt up their lower channels, and it is almost impossible to prevent silting indefinitely. In many of the smaller river systems, under natural conditions, there is very nearly a balance between the sand and mud carried into the main stream by the tributaries and the amount swept out to sea by the major stream. But in the large basins of the big rivers, the master stream receives more material than it can flush out.

River beds are built up as sediment accumulates, the streams go overbank easily when a flood comes, they tend to change their courses at frequent intervals and to deposit a blanket of sediment across the lower parts of their valleys. Borings into the lower valley of the Mississippi show that hundreds of feet of sediment have accumulated over the past several thousand years. It is quite certain, therefore, that the Mississippi, if left alone, would continue the process.

As a river mixes with the sea its current is stilled. Mud and sand carried by the stream settle at the river's mouth until a delta is built out into the sea. The effect of a delta is to lengthen the river's course, lower the average slope of the river bed, and make the current less swift. As the delta grows the stream is not capable of moving the mud from its tributaries all the way out to sea, so gradually the whole lower part of the main stream valley is built up by silting.

A river builds its delta much as a bulldozer spreads sand from a pile in one corner over the entire surface of a field. It pokes out a long finger of mud and sand, lengthens its course, slows its speed. Then it discovers that it is more efficient to start over again from its mouth and develop another finger. The second channel becomes abandoned in turn, and the process continues until a fan-shaped deposit is formed, with the original stream mouth at the apex of the fan.

River deltas are formed within the scale of human rather than geologic time. Cairo was founded as a seaport on the Mediterranean; now the Nile has separated Cairo from the Sea by eighty miles of mud and sand.

New Orleans, like Cairo, is many miles from the sea on the present main channel of the Mississippi. But the Mississippi, left to itself, would prefer to flow down the Atchafalaya channel, which is a shorter, steeper route to the Gulf of Mexico. Every bit of sand and mud the river deposits below New Orleans today encourages a change of course to the Atchafalaya channel. If this were to happen, New Orleans would no longer be a seaport, for the Atchafalaya splits from the present main river above New Orleans. The Army Corps of Engineers has the endless job of straightening the present channel, dredging sand from it, keeping it as narrow as possible to increase the current. Several hundred million tons of sand and silt pass New Orleans every year; it seems inevitable that nature will eventually overcome the efforts of man and force the Mississippi to change its lower course.

As has been indicated before, most methods of management are designed to keep rivers from "doing their thing." Their courses are shortened, channels are dredged, artificial levees are built on top of the natural ones in an attempt to keep the rivers in a controlled course, to keep their currents speeded up, and to help them carry their burdens. The rivers usually insist on building up their beds anyway, so the artificial levees grow higher and the damage from floods increases.

In planning river control it is the biggest flood that is the important one, and it requires a long span of observation to know how rampant a river may become. Floods seem to come in many different cycles; we hear of five year floods and twenty-five year floods and one hundred year floods. On the main stream of a large river the height of a flood is controlled by a combination of many factors. When they are all favorable for flooding, previous records may be surpassed. If, for example, heavy rains in the headwaters of the Missouri produce a flood crest that moves down just in time to catch one from the Ohio River, it means compounded trouble. Someday crests from the Missouri, Ohio, Illinois, Arkansas and Red rivers may coincide. Catastrophe! One aspect of planning is to build dams and reservoirs on the tributaries so that the timing of floods can be controlled and destructive coincidences can be thwarted. Dams and reservoirs can be used for power generation. The reservoirs help to settle suspended mud and sand, thus clearing the downstream water and helping to prevent the master stream from building up its bed. There is much more in favor of damming streams—lakes can be created for recreation and irrigation and stream flow can be made more nearly continuous throughout the year by feeding from the reservoir. The Tennessee Valley Authority has shown that all these factors can be successfully coordinated. A similar project at Lake Meade, behind the Hoover Dam on the Colorado River, has generated a spectacular oasis in the western desert and has provided multiple benefits.

Dams have one drawback that is an always-fatal creeping paralysis—the lakes created behind them fill with sediment. Depending upon the specific conditions, this may take twenty-five years or four hundred years, but the life of reservoirs is finite. The average lifetime of dams is so short that most dams built today will be silted up in time to cause serious problems for our great grandchildren. No one has yet figured out an economic way of removing the accumulated sediment to restore the original capacity of the reservoir.

Dams cannot be installed just anywhere. Where a stream flows in a wide valley (inevitably the site of towns and cities) the cost of blocking the entire valley is prohibitive. Valuable agricultural, grazing, or scenic areas, as well as industrial centers, may be put under water. Most proposed dams therefore raise tremendous controversies as to whether their basic long-term economics produce profit or loss. Is the cost of the dam and its upkeep balanced by the value of the electric power produced and the flood control and recreational benefits that result? Will the lowering of downstream flow during the filling of the reservoir be tolerable to water users? Will there be a market for irrigation water or electric power produced?

The future of the world's rivers is certainly predictable in general terms. They will continue to be the chief source of fresh water. Withdrawals will increase until all water reaching the sea from rivers will have been removed and returned at least once. If irrigation can be controlled, water use, even on the greatly expanded scale of the future, will not necessarily diminish total stream flow significantly. The goal will be to minimize degradation of water quality as a result of use. The cost will be great and success can come only from highly organized control of water use. The political implications are clear. If a river is to be managed and if its water is to be usable, there can be no division of its flow into autonomous political units. Administration will have to come from a single political unit whose authority encompasses the entire drainage area.

The prediction that water use in manufacturing will have the greatest relative increase in the next thirty years means that the major test for the well-watered areas will be how well they keep industrial contaminants out of rivers. It seems impossible to assess the effects of the increasing variety of waste products. To determine whether or not a given substance will have a deleterious effect takes so long, and the number of possible consequences to be considered is so great, that the only safe practice will be to set arbitrary and extremely low limits on addition of all foreign substances. The burden will then be on the waste producer to demonstrate the true tolerance levels for the wastes he is adding. A tremendous expansion of present day water treatment must inevitably take place.

In dry regions, rivers will be completely consumed. The competition will be between large scale desalination and the transport of water from distant sources. It looks as if irrigation will continue on roughly the present scale, but with a gradual change away from staple crops toward high value ones in the areas where agriculture depends upon irrigation. Better transportation will permit increasing specialization of agricultural land use. Food will be grown in the most advantageous locations and transported to the rest of the country. Manufacturing will undoubtedly change in a similar fashion. Major food production will be in the well-watered areas, with the less well-watered regions producing high value crops by intensive water use. Economics will be dictated by innovation and invention. While rivers will continue to be used, the use of their water will change drastically.

〰〰〰〰〰〰〰〰〰〰〰〰〰〰〰〰〰〰〰〰〰〰〰〰〰〰〰

CHAPTER SIX

Lakes

IN THE PARTS of the world where precipitation exceeds evaporation, a well dug deep into the ground fills with water. The level to which it fills may be a hundred feet below the land surface or it may be only inches. On the average, the depth to the water surface is twenty feet. Above the water surface of the well the surrounding soil and rock is moist; below the surface the pores of the rock are filled with water. (Beneath the land surface there is another surface below which the pores of rocks are saturated with water.) Any hole dug below this saturated level will fill with water. Lakes can be thought of as super-wells—natural depressions below the surface of the zone of saturation that have filled with water. Their origins are as diverse as are the origins of depressions themselves. Lake basins may have been scooped out by glaciers or they may be depressions in the hummocky deposits left by retreating glaciers; they may form where sections of land have sunk along fractures in the earth's crust, or in the collapsed cone of a volcano, or where a river channel has been abandoned by a stream, or where a landslide has blocked a river. Many lakes are simply expanded rivers—they have an inlet and an outlet and are kept full by the river's flow and by percolation of groundwater; others have no inlet or outlet but are fed and drained entirely by the flow of groundwater into them on one side and out on the other.

Estimates of the total volume of the lakes of the world are quite variable. The Water Resources Committee of the National Academy of Sciences-National Research Council gives the volume as 100 billion acre feet. Translated into cubic feet this is 4,356 trillion; as gallons it is 32,600 trillion; divided among the people of the world it is about 10 million gallons each. If used by the United States alone, the lakes of the whole earth would last about two hundred years at the present rate of water withdrawal.

The United States and Canada share the Great Lakes and thus have an inordinate share of lake water—a volume of 900 trillion cubic feet, or about 20 percent of the world total. The Great Lakes—Superior, Michigan, Huron, Erie, and Ontario—make an irregular chain across a thousand miles of the eastern central part of North America. They are unique in their size and continuity. The total area of the lakes is almost 100,000 square miles, and they have an average depth of 310 feet. Superior, giant of the lakes, covers almost one third of the total area, is 412 miles long, has an average depth of 475 feet and a maximum depth of 1,000 feet. It is also the favored lake, if by that we mean it lies far enough to the north to be away from population centers, is in a basin of hard rocks that add little dissolved solid material to its water, and generally retains the characteristics with which it was born. Huron and Michigan are about the same size, each about two thirds the size of Lake Superior, but Michigan is deeper, averaging 325 feet to Huron's 250. Huron, like Superior, is fairly well isolated from the hurly-burly, but Michigan, extending in a giant thumb between the states of Michigan on one side and Wisconsin and Illinois on the other, stretches down to the city of Chicago. Its shores are the sites of some of the densest population of the United States. Superior and Michigan drain jointly into Lake Huron. Michigan has an alternate outlet down the Chicago River and eventually to the Mississippi, but the volume of water that goes that route is miniscule.

Lakes Erie and Ontario are the dwarfs of the chain. Erie covers ten thousand square miles; Ontario covers about seven thousand. Together they cover only half the area of Lake Superior. But there is an important difference between them; Ontario has an average depth of three hundred feet, Erie only seventy. Ontario's average depth is greater than Erie's maximum.

Lakes Michigan, Erie, and Ontario lie in a region that is shown on maps of Canada and the United States as at, or near, the peak for the two nations in population density, power production, heavy and light industry, as well as being a center of agricultural and livestock production. These three lakes, therefore, have been subjected to pressures unmatched almost anywhere else. Of the three, Erie has suffered most, chiefly because it has by far the least water volume—only nineteen and a half trillion cubic feet, as compared to Ontario's sixty-one trillion and Michigan's two hundred and three trillion. In many ways, Erie is unique among the Great Lakes. Its seventy-foot average depth makes it a shallow pond when compared to the several hundred foot depths of all the others. If the Mississippi River could be diverted into the Great Lakes, it would take only one year to replace the water of Lake Erie, but almost ten years to wash out Lake Michigan.

Some idea of the size of the Great Lakes can be had by imagining them as the only source of water in the United States. Without any renewal from rain or groundwater percolation, they would last forty-one years at the present rate of use. The water in the lakes comes from a small local land drainage area, plus the excess of precipitation over evaporation on the lakes themselves. Their year to year fluctuations in level are largely controlled by the balance of precipitation and evaporation; drought periods drop the levels, wet years raise them. The fluctuation is a number of feet, enough to have severe effects at either high or low water. When lake levels are down, ships scrape the bottom of the channels dredged in the rivers that connect the lakes; when levels are high, storm waves pound across the beaches to flood low shores and eat away high-priced real estate.

The waters of the lakes, before the influence of man, were of excellent quality throughout. Lake Superior, at the head of the chain in its hard rock basin, is still nearly as pure as the Amazon River. Michigan's near shore waters and shallow subsidiary bays, like Green Bay, Wisconsin, have suffered, but as a whole the lake is in good condition still. The lower lakes, Erie and Ontario, are in basins that are very limey compared to the Superior basin and have naturally higher concentrations of the hard water constituents, calcium and magnesium. They also are the recipients of dissolved materials that flow in from the upper lakes.

During the summer, the surface waters of lakes are heated and become less dense and so float on the colder water. The lakes are then stable in the sense that there is no density drive to overturn the waters and circulation from bottom to top is brought about only by forces like a strong wind that blows the warm surface waters to one side of the lake, exposing cold, deep water on the windward side. Such conditions, although they do not cause mixing throughout the lake, can be important to anyone living on the shore.

The effects are quite striking on Lake Michigan. In Chicago there is a strong belief that on hot days the water feels freezing cold only by contrast with the furnace blast of the air. The facts are that when it is really hot in Chicago the waters of Lake Michigan at the Chicago beaches are much colder than normal. When the temperature climbs into the nineties in Chicago the heat wave usually rides in on the wings of a strong, dry, steady southwest wind. The wind blows the surface water onto the eastern shore of the lake and deep, icy water moves upward and westward to take its place on the beaches. The greater the heat wave the stronger and more persistent are the winds, and the deeper and colder the water that slides up onto the beaches. In the early stages of the development of this circulation, swimmers encounter long fingers of cold water sneaking up from the depths and within a few feet they may swim from seventy to fifty degree water. In the meantime, the swimmers on the eastern shore, some sixty miles away, are delighted, for the water is refreshingly cool at seventy-odd degrees, whereas the Chicagoans are either roasting on the beach or freezing in the lake. Nor is there any happy medium—comfort does not consist of ice water to the waist and a blast furnace above.

The winter air temperature in Chicago drops sharply, well below that of the deepest coldest waters. The surface water becomes colder and denser than the fifty degree deep water and sinks downward, displacing upward the waters that have been isolated throughout the summer. The lake overturns continuously and the average water temperatures get lower and lower. Cooling and mixing take place at such a rate that it seems as though the whole lake would arrive at freezing temperatures at the same time and end up as a massive block of ice. But water, the strangest of all liquids, has a most peculiar property. It becomes denser as it becomes cooler, until it reaches 39° F (seven degree above freezing); then it starts to expand again. This reversal comes just in time to prevent solidification of lakes and rivers from top to bottom. After cooling to thirty-nine degrees the surface waters are less dense and float instead of sink. They continue to cool and float until they are frozen. Ice floats because it is less dense than water; thus freezing of lakes takes place from the top down.

Lake Michigan is so deep that even during severe Chicago winters there is not time enough to cool and overturn all the water to the thirty-nine degree temperature of maximum density. The lake never freezes over entirely, although extensive ice packs may be generated in shallow areas.

The behavior of Lake Michigan is typical of what we could say is "proper" for a lake. The deep waters, although isolated from the atmosphere during the summer, manage to retain some of their dissolved oxygen throughout their confinement. There is enough oxygen to react with organic materials that settle into the deeps, or to take care of other oxygen consumers like the iron in some of the minerals. Fish can swim where they will, bacteria cannot thrive, the waters are pure, sweet-smelling and drinkable. The sediments that accumulate on the bottom of the lake are well oxidized before they are covered by new sediment, so that little further bacterial decomposition, with the accompanying generation of foul gases, can take place after they are buried.

This delightful situation lasts as long as the oxygen content of the waters exceeds the demands placed on it by oxygen consumers—chiefly organic materials. But if a time comes when, for whatever reason, the dissolved oxygen is used up, the whole character of the lake changes, and eutrophication (that widely publicized word) sets in. Organic material, upon which bacteria feed, accumulates on the bottom or remains suspended in the water. The anaerobic bacteria—those that live in the absence of oxygen—are happy and set about multiplying and producing their waste products, among which are hydrogen sulfide, methane, and ammonia, plus a host of other noisome substances. The entire initial fauna of the lake is wiped out in the anaerobic zone. Fish cannot enter it, burrowing animals die, and the water, at best, is unpleasant for human consumption.

Eutrophication is not by any means an unusual situation created by man. The natural oxygen balance is positive in some lakes and negative in others. Considering that a tiny pinch of sugar fermented in a quart of water is enough to use up all the dissolved oxygen normally present in the water, it should be obvious that it does not take much organic material to use up all the dissolved oxygen in lakes. In shallow lakes wave action alone may keep the water oxygenated. If water weeds begin to grow profusely two things result: the weeds still the wave action; when the weeds die they consume oxygen. Eutrophication results naturally under these circumstances.

Man causes eutrophication chiefly in two ways. He may directly add to the water enough organic material of almost any kind—sewage, garbage, leaves and grass cuttings, pea pods or apple skins from food processing factories, etc.—to overwhelm the oxygen supply. More commonly, he adds chemical nutrients that stimulate the growth of aquatic plants which then do the job of eutrophication in a natural way. Best known of these additives is phosphorus, found in detergents, toothpaste, raw sewage, and also the effluent of most sewage treatment plants. The difficulty with phosphorus is that it is so extremely effective in stimulating plant growth that even small quantities can cause difficult problems. Ordinarily its concentration in natural waters is small enough to be measured in parts per billion. It contributes little to the tissues of the organisms that use it, but it is essential for growth. For aquatic plants it acts as a fertilizer in the same way that fertilizers stimulate growth of land plants. In effect, many lakes have been fertilized, thereby increasing their organic productivity. Nitrates also "fertilize" lakes. Water draining fertilized fields contributes nitrates. It is a grimly amusing fact that the furor over phosphate-containing detergents spurred the development of phosphate-free detergents containing nitrates. These compounds, when discharged to lakes, are acted upon by bacteria and nourish the aquatic vegetation even better than does phosphorus. In a sense, fertilization by phosphorus or nitrogen could be considered a biologic "good," and has even been suggested as a way to increase food production. It happens that most people want lakes and rivers that are on the sterile side, with lots of "wasted" oxygen in the water. If we think of lakes as people and of organic matter as their food, our best friends are lean and hungry lakes that pounce on every scrap of nourishment they can get. Feed them too much and they become sluggish, bilious, and dyspeptic.

Some interesting experiments have been made to reverse the eutrophication of small lakes by aerating them directly or by forcing them to overturn by circulating the deep stagnant waters with a pump. The power required to circulate the water is small; only the slight difference in density between the top and bottom waters need be overcome.

Waters that have lost their oxygen are not necessarily carriers of disease and can often be drunk safely if not pleasantly. The great danger is that eutrophic lakes provide good environments for the growth and spread of bacteria. Persons infected with typhoid, cholera, amoebic dysentery, and other diseases, pass on their anaerobic germs to the sewage. Ordinarily the germs have a hard time surviving in air or aerated water, but if even a few manage to enter the deoxygenated waters of an organic-rich lake they can flourish and reach infectious levels throughout a large area.

Lake Erie, shallowest of the Great Lakes, has suffered most from man's activities. There are many coincidental factors that have combined to speed Erie's decline. With a much smaller water volume than the other lakes, it is more sensitive to having its composition changed. It inherits salts that have been added by man to Superior, Michigan, and Huron. It is nearly surrounded by a densely populated area that is heavily industrialized. It is fed by the Detroit River, which, as it flows some fifteen or twenty miles to Lake Erie, first passes the city of Detroit on one side and the city of Windsor on the other. Then it flows past the Ford River Rouge plant, one of the largest manufacturing plants in the world, plus several steel plants. As if that were not enough, a great chain of basic chemical producers sit on its banks, their operations based on the salt deposits that underlie the area: Wyandotte Chemicals, Solvay, Pennsylvania Salt Co., the Brunner Mond plant. Chemical wastes are stored on islands in the river. Five hundred foot freighters carrying iron ore from Lake Superior to the steel plants, coal from Ohio, automobiles from Detroit to Ohio, or bringing oil from across the Atlantic, form a continuous procession. Accidental or deliberate discharges from any of these sources contribute many undesirable materials to the river.

Great efforts are being made (and with considerable success) to reduce the additions of a complex set of substances to the Detroit River, but still the water delivered to Lake Erie reveals human influence on every element present. As the river empties into Lake Erie the water moves past the city of Toledo at the west end of the lake, then has to run the gamut past Cleveland, sitting midway on the southern shore, to arrive at Buffalo at the east end.

The near-shore zones of the lake, where mixing of the pollutional additions from a million sources takes place, have suffered most. The shallow areas have been overwhelmed. The whole lake teeters on the edge of complete oxygen loss during the summer months, when the warm surface waters do not mix with the colder waters below.

Just what would happen if all pollution could be stopped tomorrow is still a subject of much study. Simply because Erie is small the average renewal time of its waters is only a few years. At first glance one would say that complete rejuvenation should not take much longer. Unfortunately, it is not yet known how long the effects of the organic materials that already have accumulated will last, or how long complete flushing of the lake, including isolated shallow areas, will take. Eutrophication tends to be irreversible. One of the important reasons is due to the phosphorus that accumulates in the bottom sediments and can be released back to the water. Organic production can continue to overwhelm the oxygen supply. Once the biology of the lake has been drastically altered, can the original organisms return? Will it be necessary to "manage" the lake, to reseed it, perhaps successively, with different plants and animals? A group of scientists plan to pollute a small, clean lake deliberately in order to watch exactly what changes occur and to discover, hopefully, how to reverse them. It seems hard to believe that such a drastic step is required to learn to undo the harm that has been done.

There are but a handful of other lakes that rival the Great Lakes in size; their closest competitor is the chain of lakes in central Africa. Lake Nyasa has a length of three hundred and fifty miles and a maximum depth of twenty-six hundred feet, twice that of Lake Superior, but its volume is slightly less—396 trillion cubic feet to Superior's 413 trillion. Lake Tanganyika is longer than Superior and has a volume of 283 trillion cubic feet. The origin of the African lakes, with their tropical location, of course has nothing to do with glaciation; they lie in a great sunken valley that runs north and south through Africa, splitting the continent into two parts.

Deepest of all the major lakes is Lake Baikal, deep in central Asia. Shorter than Superior by a hundred miles, it is a little more than a mile deep—five times deeper than Superior. Only one third the area of Superior, Baikal contains almost two thirds as much water. The Russians are having a difficult time with pollution; they, like us, were not aware of the effects of nutrient additions to lakes until major changes in Lake Baikal had already occurred.

Highest of the major lakes of the world is Lake Titicaca, three miles above sea level in the high plateaus of Bolivia and Peru.

Fame does not accrue to lakes through sheer size: Loch Ness, home of the legendary monster, covers but twenty square miles. It is, nonetheless, formidably deep and there is plenty of room for "Nessie" in the deep pockets that reach down more than seven hundred feet.

Lake waters tend to be either quite fresh and dilute, or very salty indeed. When lakes lie in areas with an excess of precipitation over evaporation, the streams and groundwaters that feed them are good quality waters and the lakes themselves receive an excess of rainwater poured on their surfaces every year so that they remain fresh. In arid areas, especially those with undrained depressions, lakes have a hard time surviving. The Great Salt Lake of Utah is a fine example. Its water today is eight times as salty as sea water, with 26 percent dissolved salts. Once the Great Salt Lake was deep and fresh. It was fed by abundant rain during glacial times and had an outlet to the west. When the glaciers retreated, the climate grew arid and the level of the lake fell below that of its outlet. The complicated story of higher lake levels is written on the walls of the Wasatch mountains that rise behind Salt Lake City at the east end of the lake. Several "benches," at different levels above the present lake, have been cut into the mountainsides by the water. Today the lake is shallow and its inflow is barely enough to cope with the high evaporation of the Utah desert. Salt has been produced at the western end of the lake for many years, but with low yield. More recently the chemical industry has taken a greater interest in the concentrated brines and there has been a great deal of discussion about possible long-term effects on the lake if salt production is increased. One of the most unusual features of the Great Salt Lake is that its composition is nearly that of sea water that has been evaporated to one eighth its initial volume. Anyone who did not know the lake history and had available only modern chemical analyses of the water would deduce that the lake had once been an arm of the sea, since cut off, and concentrated by evaporation to its present salinity. Other lakes in the Great Basin country of the United States are equally briny but none of them has a composition like that of sea water.

Natural chemical processes in the Great Salt Lake are complicated by the hot summers and cold winters of Utah; in winter the drop in temperature causes precipitation of salts like sodium sulfate, which redissolve the succeeding summer. The lake is so briny that swimming is more an experience than a recreation—swallowing a mouthful of water is eight times more unpleasant than swallowing sea water. Because the lake water is 20 percent denser than fresh water a swimmer floats so high that he almost feels he has to reach down to swim.

Lakes are called *closed* if water leaves them only by evaporation and *open* if they have an outlet. Closed lakes are obvious candidates for becoming saline. There are many small salt lakes in the western desert in closed basins between the mountains that have no outlet to the sea. There are playa lakes, which may cover many square miles of flats with a foot or two of water after the spring rains run down from the surrounding mountains, then evaporate to dryness during the summer, leaving a dazzling white plain of salt deposits. Many of the saline western lakes have filled and shrunk markedly within historic time. Lake Winnemucca in Nevada is now dry. In 1840 it was also dry, but between 1867 and 1882 it contained water to a depth of fifty feet. It has been as deep as eighty-seven feet and has occupied as many as one hundred and eighty square miles, yet since 1945 it has been dry. Whether it will again fill with water is unknown and unpredictable.

A few lake basins have enough drainage into them to overcome the high evaporation rate of the desert, so they manage to keep some water throughout the year. Their compositions are varied and take their character from the nature of the rocks of the adjacent mountains. The lakes of the eastern California basins are high in boron because of the volcanic nature of much of the surrounding terrain. At Kramer, California, a thick deposit of borax, that was deposited in a lake millions of years ago and was subsequently covered over by thick layers of sand and gravel, is being mined. Today many of the existing salt lakes, or fossil salt lakes as at Kramer, are the sites of major chemical plants busily separating out the salts deposited by the action of the dry desert air.

Lakes, like artificial reservoirs, can become extinct by filling with sediment. The state of Minnesota has thousands upon thousands of lakes formed some ten thousand years ago by glacier retreat. Two kinds of lakes were formed. Some are hollows in the irregular deposits of debris from the melting ice sheet, and others, farther to the north, are basins that the ice gouged out of bedrock. Lake Superior, with its thousand feet deep, four hundred mile long basin, is an extreme example of bedrock gouging. Most of the other bedrock lakes are small, from a fraction of a mile to a few miles long, and with an average depth of less than a hundred feet. Many are not closed basins—water percolates into them from the soils and percolates out the same way. Plants grow in the shallow water, trapping mud that is washed into the lake. Soon the shallow water zone is filled with a mixture of mud and organic remains of plants and becomes swampy. Slowly, typical land vegetation encroaches; bushes and trees migrate outward from the original shore as filling continues. Many shallow lakes, once a mile or so wide, have thus been completely filled and have rejoined the land.

Small bays on large lakes can become extinct in the same general manner, or they may fill in with sediment that has been swept into the bay mouth by waves and currents and dropped in the slack water of the bay, which then affords a foothold for water weeds. Such bays are highly sensitive to additions of phosphorus. Under natural conditions, the waves may keep the beaches clean and well washed, but, if the plants that grow in the mud beyond the beach and nearshore zone become too prolific, they not only form a thick organic muck that is distasteful to the swimmer, they also cut off the waves. Gradually the beach is invaded by weeds, grass, and bushes from the land. Eventually the water's edge is marked by a tangle of land vegetation on one side and water vegetation on the other. All that is left of the beach is a sandy soil.

Lakes are not permanent features of the landscape. They form and disappear through time. A small pond may be the remnant of a large lake that is drying up, or it may be just an early stage in the growth of a large lake. The geologically ancient Lake Bonneville which once covered twenty thousand square miles in Utah is now Great Salt Lake, a tiny remnant of the ice age lake.

In the Midwest where there is abundant groundwater, many man-made lakes have been created. It was found that any deep hole dug into the ground would fill with water by percolation from the sides of the excavation. Many adult residents of the Midwest remember learning to swim in abandoned quarries. Today there are numerous suburban housing projects that must be managed by quarry-taught swimmers, for they are built around artificial lakes. If lakes are dug in sand and gravel, the excavated materials can be sold for construction. Houses are built on the lake shore and in no time at all the water is congested with happy swimmers and boaters.

Lake creation by simple excavation stops about at the western edge of Illinois and Wisconsin, where it is limited by the critical line of equal evaporation and precipitation. West of the line an excavated lake tends to become salty in a relatively short time. The natural lakes of western Minnesota, for example, are almost all very salty.

To combat the consumption of water by evaporation from lakes and reservoirs, various ways of protecting the water surface have been devised. The most successful method to date is the application of a wax-like organic compound (a so-called long chain or higher alcohol) which spreads over the water surface in a layer only one molecule thick. If the film can be maintained unbroken over the entire surface, it may reduce evaporation by as much as 70 percent. Only one fourth of an ounce of the compound is needed for an acre of water surface. On small bodies of water it can be distributed by boat; for large areas a plane is required. Tests in the United States and Australia show that under actual conditions, evaporation can be cut by one third to one half. Wind, waves, and boats are the chief foes of the protective film. The cost of water saved in this way is about ten cents a thousand gallons, or about one tenth of the current cost of desalination. The chief cost of maintaining the film is man power and boat time to repair broken patches. There is a certain amount of worry about the long-term effects of such a film on the water surface. It may well be that its use will be restricted to reservoirs in which the chief aim is to preserve water, not plant and animal life. Some scientists fear that the film, although it is not toxic to animals, may reduce oxygen exchange between water and atmosphere, or may become detrimentally concentrated in plant or animal tissues.

Lakes do not figure largely in the total fresh water supply of the world, but they do have great local importance. Most of all they are a highly significant part of what is coming to be called the quality of the environment. The desire to live beside or near water is strong in many people; the value of lake front property in the United States has grown at a tremendous rate, especially in the last few years. Undeveloped frontage is vanishing rapidly. Recreational use of lakes and reservoirs has reached the saturation limit in many places; sail and motor boats are so numerous that traffic problems have developed.

Lakes must be managed to survive. Additions of waste materials to them must be restricted, the amount of water that can be withdrawn from them must be controlled. The Wildlife and

Fisheries people must work to keep the lakes stocked with species that can survive the water and habitat changes that are inevitable as the struggle to prevent pollution goes on. At the moment it is touch and go for many lakes, and it is an open question whether our understanding of lake processes, and application of restorative measures, can save them for our use.

~~~~~~

CHAPTER SEVEN

The Unseen Ocean

Exploration *by* the sea
is far more ancient than exploration *of* the sea. Only a hundred
years ago was a systematic study of the ocean started. Although
the sea had captured the imagination and curiosity of men for
thousands of years, prior to the nineteenth century study of the
oceans was haphazard, a by-product of voyages with other purposes. Knowledge of the sea was mostly an accumulation of the
lore of the sailors and fishermen. In 1868 Professor Charles
Wyville Thomson, studying biological conditions in the deep
sea off the coast of Scotland, unexpectedly found animal life at
great depths. Much public interest was aroused, and several
scientific expeditions were launched throughout the 1870s, a
rich decade for oceanography. The *Challenger* expedition, directed by Thomson, explored the Atlantic and Pacific during the
years from 1872 to 1876 and is still looked to as a model of
excellence in scientific work. Its results, including water analyses,
weather reports, and magnetic, geologic, and botanical observa-

tions, were published in fifty volumes. The pace of oceanic exploration has accelerated remarkably since that voyage, especially in the past decade.

Mapping of the ocean floor using the reflection of sound waves to measure depth has been the single most revealing kind of exploration. Before development of the sonic method, which continuously records depths as fast as the ship can cruise, it was necessary to throw a weight attached to a wire overboard, let it sink to the sea bottom several miles below, then winch it back up. Hours were spent to get a single depth measurement. With the sparse data available from such manual soundings, the ocean floor seemed to be flat, but with sonic profiling the concept of flat-floored oceans was quickly changed. There are indeed extensive abyssal plains, but there are also submarine mountain chains, great rifts and fractures, trenches reaching down more than six miles, mountains far below the surface yet somehow decapitated, troughs partially filled with cascaded debris. The structures of the ocean floor demand attention in detail nearly comparable to that required to decipher the geologic complexities of the continents. Second only to the revolution in our thinking about topography and structure of the ocean floor has been the demonstration of abundant life in the deeps.

Oddly enough, man's descent to the deepest place on earth went almost unnoticed in comparison to the analogous feats of scaling Mt. Everest or reaching the North Pole. Why these events captured the public imagination but the conquest of the deeps did not is a mystery. Who remembers today that Jacques Picard and Don Walsh descended 35,600 feet in the bathyscaphe *Trieste* to the bottom of the deepest hole in the oceans in the Marianas Trench in January, 1960? It may be that because their feat was so sudden and unexpected and not followed by many more deep dives and more publicity it has almost escaped notice.

Now that there are extensive data, some of the overall properties of the seas are known. If the world ocean is divided into the Pacific, the Atlantic, and the Indian, half of the water is in the

Pacific, and a quarter each in the Atlantic and the Indian. The world oceans hold eighty percent of the world's water, some 370 billion billion gallons! Even though the volume of the oceans is unimportant on the celestial scale, a canal built from earth to the moon and filled with the earth's ocean water would have to be 150 miles wide and ten miles deep for its entire length. The well-lighted ocean surface obscures the fact that the great body of the oceans is black and cold, with an average temperature of three to six degrees above freezing. Light penetrates at most only a few hundred feet below the surface; temperature diminishes quickly with depth.

The average density of world oceans is 2.754 percent greater than that of distilled water. The dissolved salts are 3.47 percent by weight, or 34.7 parts per thousand (‰) in the terminology of the oceanographer. The Pacific, the Atlantic, and the Indian oceans all have variations in properties from top to bottom and from north to south and east to west. It has taken thousands and thousands of measurements of temperature and salinity to establish that the average, the *common*, water of each is really a little different from that of the others.

Atlantic and Indian waters average about half a degree warmer than those of the Pacific; the Atlantic is a little more saline than the Indian, and the Indian a little more saline than the Pacific. Yet the differences are only a matter of a fraction of a degree and a fraction of a percent and are far overshadowed by variations within any one ocean. The deep waters of all the oceans are alike; the big differences come near the shores and in the upper half mile of the open sea, and perhaps next to the bottom. The Atlantic shores of North and South America are fed by large rivers such as the Paraguay, the Amazon, the Orinoco, the Hudson, and the St. Lawrence, whose waters influence the near shore ocean water. For many miles off Cape Cod the salinity may be as low as 32‰ as opposed to the oceanic average salinity of 35‰. Surface waters in the Arctic and Antarctic regions of the Atlantic and Pacific are diluted by fresh water from melting ice and their salinity may be down

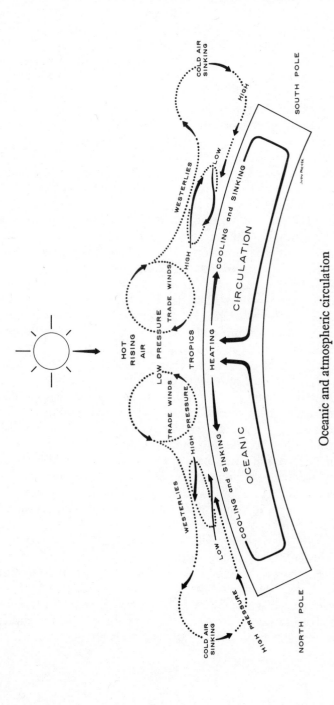

Oceanic and atmospheric circulation

to 33‰. At the equator, salinity of surface water is average, for the water evaporated is returned to the sea. At about twenty degrees North and South latitudes, the dry trade winds evaporate water and transport it away, so that the salinity of the oceans there may climb to 37‰.

Of the average 34.7 parts per thousand salts, sodium and chlorine, the ingredients of common salt, make up almost thirty parts, while magnesium and sulfate add four more. Of the other one part, calcium and potassium add four tenths each; carbonate about one tenth. Thus only eight elements—oxygen and sulfur, chlorine, sodium, magnesium, calcium, potassium and carbon— make up 99 percent of the sea salts. But in that last 1 percent are phosphorus, nitrogen, and silicon, the chief nutrients for the microorganisms that are the base of the oceanic food chain. There are surely also traces of every one of the ninety-two naturally occurring elements; sixty have been detected so far.

Even where sea water is diluted, or where the salinity is increased by evaporation, the proportions of the major elements are essentially unchanged. Not until recently has it been possible to analyze sea water with an accuracy great enough to prove that there are indeed slight variations from place to place in the ratios of the major elements. The dynamic daily chemical changes in the oceans involve the minor elements; the major ones provide the stage setting. An analyst working with a gallon or two of sea water is delighted to detect an element like titanium that occurs as only one part per billion of the sample, yet one part per billion adds up to four and a half tons in a cubic mile of sea water.

The tremendous amount of water in the oceans and its availability has always challenged chemists and engineers to develop processes to extract the elements. They dream of huge installations beside the sea that will supply all the world's chemical requirements. They know that a million gallons of sea water could yield 40 pounds of boron, 570 pounds of bromine, 11 pounds of fluorine, a pound of rubidium, and so on. A million gallons is no more than some desalination plants are now treating each day to get fresh water—why not somehow combine the operations? Some of the basic economics may tell why not. It costs about a thousand dollars to produce a million gallons of fresh water from sea water; the value of the 120 tons of salts remaining would be several thousand dollars, but the cost of separating, refining, packaging and marketing would lower the profit considerably. Current processing costs are prohibitive, except for a few high-value elements.

Everyone wants to get gold from sea water. At one time it seemed an attractive project, with good prospects of becoming a commercial success. But the gold analyses on which the cost estimates were made were in error. When recovery was attempted, the predicted amount of gold wasn't there. So far the total recovery of gold from sea water is valued at less than a tenth of a cent, and the trace amount in the ocean is of the order of a penny's worth per million gallons. Only about fifteen elements occur in sea water in concentrations high enough to permit extraction of more than a dollar's worth from a million gallons. In addition to the major elements already given, they are bromine ($200 worth from a million gallons of sea water), boron ($3), strontium ($2), nitrogen ($1), lithium ($36), iodine ($1), and iridium ($4). It is quite unromantic to find that magnesium ($4,000) and plain old sodium chloride (table salt) ($900), are much more interesting financially than silver (2¢) or zinc (1¢). As might be expected from these figures, bromine, magnesium, and common salt are the only substances currently being commercially extracted. Still, it is comforting to know that even at these low concentrations, there are sixteen thousand billion dollars' worth of gold and silver in the world ocean—enough to finance even the United States for a generation or so.

Since ancient times sea water has been evaporated to get common salt. If it is evaporated to dryness, the total sea salt left behind has enough magnesium sulfate (Epsom salts) in it to be unpleasant. Also, the mixture of salts tends to cake and to be moist and sticky. To get table salt, sea water is passed through a series of shallow ponds. In the first ponds, calcium carbonate and calcium sulfate (gypsum) precipitate. When the sea water has been reduced to a tenth of its original volume it is moved into a new pond and nearly pure sodium chloride is removed. If the volume is reduced to about one fiftieth, potassium and magnesium salts—the bitterns—precipitate.

At the bittern stage it becomes nearly impossible to evaporate any more water from the brine without heating it artificially. The concentrated sea water gets thick and soupy and tends to pull moisture out of all but the very driest air, instead of giving it up. Most outdoor salt pans produce only sodium chloride; their bitterns are returned to the ocean.

From time to time in the geologic history of the earth, seas crept onto the continents. As much as 50 percent of North America was under water during the Silurian Period, some four hundred million years ago. Water circulated from the open ocean into these seas and back again. Where the climate was dry and the sea broad and shallow, salinity increased. In some places the water got back to the ocean before anything but calcium carbonate was removed, elsewhere the water reached the gypsum stage, or the salt stage of concentration. In what are now the states of Michigan and New York thick gypsum and salt deposits were formed. The continental seas behaved like a huge salt-pan operation, in which sea water was circulated onto the continents, stripped of part of its dissolved salts, then returned to the deep ocean basins. The total amount of salt and gypsum still stored in rocks of various ages is comparable to the amount in the sea today. If we could restore the salt in continental salt deposits to the ocean, salinity would be almost doubled.

Today natural removal of salt from the sea is unimportant, but the principle of the processes that formed the ancient salt deposits is illustrated by conditions on the wide shallow platform on the west side of Andros Island in the Bahamas. The sea water is only six to ten feet deep on a shelf fifty or sixty miles wide. This shelf deepens abruptly into the channel between Andros and Florida. Sea water moves onto this shelf, circulates slowly across it, taking about nine months to return to the deep ocean. During this slow circulation, evaporation exceeds precipitation and the salinity of the sea water increases by 30 percent. In this case, only calcium carbonate is removed, but it is easy to see that if the size of the shallow water area were increased several times, gypsum and salt would precipitate.

Of all the elements that are carried in dissolved form to the sea, most is known about calcium. It is the most important builder of shells. Near the shore corals, oysters, clams and many marine plants secrete skeletons of calcium carbonate. In the surface waters of the deep ocean two kinds of microorganisms take calcium from the water for their shells—the forams and the coccoliths. The calcium and carbonate contents of ordinary surface waters almost everywhere are high enough for the spontaneous formation of solid calcium carbonate without intervention of organisms. For reasons still mysterious, despite many years of research, this rarely, if ever, happens. The calcium is deposited through the agency of organisms. Because of the dependence of the bulk of the plant community on light for photosynthesis, and the dependence of animals on plants for food, almost all shell-making goes on within a few hundred feet of the surface.

Corals take advantage of the fact that their dead shells will not dissolve in the calcium-saturated surface water to build imposing reefs. The live coral animals live only on the reef surface, and are dependent upon good water circulation to bring to them the particles of organic matter suspended in water that they need for food. They also must have light, for the coral animals have tiny green plants embedded in them. The corals are dependent for their existence on the photosynthetic activities of the plants. Where there is a variety of coral species, competition for light and good water circulation gives a ready explanation for the multitudinous forms developed. There are coral "fans" with lacy open structures, brain corals with a cerebrum-like convoluted surface, horn corals with many spiky fin-

126

gers, branching corals like delicate trees; each has its way of exposing every tiny coral polyp to as much water and light as possible. It is no wonder that the surface of a reef is an open structure, with thousands of small caverns and grottoes, forming protective pockets for fish and other sea creatures, and surfaces on which weeds attach, providing feeding grounds for many types of marine life. A coral reef is a complex community of unsurpassed variety and beauty.

Corals are temperature sensitive and find waters cooler than 55–60° F are too cold to live in. Bermuda, at thirty-two degrees north, has the highest latitude full-fledged coral reefs in the world.

On steep shores fringing reefs develop, extending out into deeper and deeper water until there is an equalization between the rate of building by the combined efforts of the organisms and the rate of destruction by waves and by reef-eating and boring organisms. On the outer reef the pounding of the waves prevents the corals from building their delicate forms so they are forced to minimize their surface area and the number of productive animals. Many long stretches of shore cannot support reefs because there is too much sediment in the water. Where rivers discharge, the fine muds slowly settle out and corals become coated and die. They are so sensitive to their environment that it is hard to keep them alive in aquaria.

In the stretches of geologic time coral reefs have nearly always been more important than they are today. The total area of shallow water is now small compared to Periods like the Ordovician, some four hundred and fifty million years ago, when half of North America was covered by warm, clear, shallow seas, ideal for calcium carbonate-shelled creatures. Today most of the calcium carbonate deposition has shifted to the deep sea, where the tiny forams and coccoliths sink down through the miles of sea water to accumulate as deep sea oozes. Only the Great Barrier Reef, which stretches in a band a thousand miles long and fifty miles wide down the east coast of Australia, has any semblance to the widespread reef conditions of bygone days.

Each year some twelve hundred million tons of dissolved calcium carbonate come down the rivers and are mixed with ocean water. Each year just about the same amount is accumulated in shells of one kind or another, mostly in the deep sea. We deduce that each year, on any long term average, twelve hundred million tons of old reefs or oozes must be somehow uplifted above sea level to become the source of the shell materials of the future. If there were not this renewal by uplift, all of the calcium on the continents would have long ago been carted to the sea by rivers. But mountain-forming processes are always at work somewhere; the earth's crust is very active, especially when seen in the long perspective of geologic history. Some of the calcium now moving to the sea comes from the top of Mt. Everest, which was raised millions of years ago from beneath the sea.

Every year, many times as much calcium is consumed by organisms as is added to the oceans, but only a small fraction of what is consumed accumulates somewhere on the ocean floor. Most of the shells of the forams and coccoliths are somehow destroyed before they settle to the sea floor. They may dissolve in the cold deep waters where both the low temperature and the high pressure favor their dissolution.

The twelve hundred million tons of calcium carbonate precipitated from the oceans each year, sufficient as it is to nourish all the reefs and the open ocean plankton, is less than a millionth of the total calcium in the oceans. The amount coming in or the amount precipitated could fluctuate wildly from year to year and we could never tell the difference by analyzing sea water. It would take perhaps a thousand years of calcium being brought down by the rivers, without any removal at all by organisms, to raise the total in the oceans enough to be detectable. Yet on the geologic time scale of hundreds of millions of years, we can be quite sure that all the calcium in the oceans has been renewed many, many times. The reason for the constancy of ratios of elements in sea water becomes apparent: if all the waters mix within a few thousand years, the average calcium atom has time to circulate through all the oceans a thousand times before it is captured by a coral or a clam and then buried in the sea floor.

To predict what will happen to all the substances that go into the sea, at least as much must be known about them as is known about calcium. The life history of a calcium atom runs something like this. After hundreds of millions of years buried in the land in limestone rock, it finally is exposed at the surface and is dissolved from the rock in rain and soil water. In a few years it works its way to a stream, carried by the percolation of the water through the soil and rock. Once in a stream, it moves swiftly to the ocean in a few days or weeks. If it happens to be carried down the Hudson past New York, it oscillates its way down the lower reaches of the river where the waters of the Hudson mix with the Atlantic tides. It may take two more years to work its way across the fifty miles or so of the continental shelf to the open ocean where it truly belongs to the sea. Once in the sea there is one chance in twenty that it will escape from the sea surface in a bursting bubble and be wafted back onto the land. Nineteen times out of twenty it will wander the seas for a million years. Once each hundred thousand years or so it will be consumed by a coral and built into its shell or used by a tiny foram in the tropical seas. Yet nine times out of ten it will escape again, perhaps as the foram sinks into cold bottom waters and dissolves. Sooner or later the atom will find itself in a shell that falls to the bottom and gets covered up and so is locked up for another few hundred million years. Whether a calcium atom, once buried in the muds of the deep sea, has much chance of getting back up to the bottom of the sea is not yet known.

When a calcium atom, in its long travels, is in the surface water of the ocean it has a relatively lively time, being churned about by the waves, and can figure on only ten years before it is transferred to intermediate or deep waters. If it is mixed down into the intermediate waters it may stay a few hundred years; if transferred to the deep waters it may spend a thousand years or more moving laterally with its neighbors before transferring again. In its travels it has a good chance, during its oceanic residence, of moving into the Mediterranean in the surface currents at Gibraltar, and moving out again on the deeper counter current after about eighty years.

The histories of the other elements brought into the oceans dissolved in stream water are similar to that of calcium, but the ocean chemists haven't yet been able to trace them in comparable detail. The amount of magnesium discharged into the sea each year is known, but so far no adequate "sinks" or repositories are known that can account for removal of magnesium at the same rate it is going in. On the other hand, the chemists have a certain degree of confidence that such sinks will be found, because they know magnesium could not accumulate indefinitely without causing important and predictable changes in ocean chemistry. Because sea water has supported life like that of today for so long—hundreds of millions of years—the income of the various elements must continuously equal their outgo; otherwise the oceans would have become uninhabitable. True, magnesium could be accumulating in the oceans today, but the long record of a past during which it did not accumulate indicates that the absence of a way to get magnesium out of the ocean today is more likely because of our ignorance of oceanic processes than due to an actual current accumulation of that element.

Most of the material carried into the sea is not dissolved in the rivers but is carried in suspension in the water or rolled along the bottom of the stream. Eighteen billion tons of sand and clay are washed into the sea each year, in comparison with 4 billion tons of dissolved matter.

In contrast to the dissolved salts of the rivers that are mixed with sea water and circulate through all the oceans before they are removed, most of the material carried in suspension in the water or washed along the river beds stays close to shore. Some goes to filling up estuaries, other to building deltas. Some is swept along the shore by the action of waves and long shore and tidal currents. Sea water, owing to the electrical properties of its dissolved salts, causes most of the finest muds to coagulate and settle out faster than they would in fresh water. It is only a tiny fraction of the total river load that manages to get far out to sea and to become a part of the abyssal sediments. Heaviest deposition is on the continental shelves, but there are several mechanisms by which sands and muds can be carried out beyond the shelves onto the continental slopes or even into the deeps.

Strong currents can carry debris from shallow water into deep water. Sediment moves down submarine canyons to pile up at the bottom of the continental slope. Some of the canyons were cut during glacial times, when the sea was 400 feet lower than today and most of the present shelves were dry land. But others must have been carved below sea level by sediment-laden waters, denser than ordinary sea water because of their suspended mud. No one has actually seen one of these turbidity currents in action in nature, but there is evidence of their activities. The walls of many of the canyons are scoured clean. The debris has formed submarine deltas at the canyon mouths.

Experiments in the laboratory confirm the natural process: if a muddy suspension is poured into the upper end of a water trough with a gently inclined bottom, the muddy water will move at high speed down the bottom of the trough, maintaining its identity as a separate layer for long distances. As the current subsides, coarser particles drop out first, followed by finer and finer ones, making characteristic, widespread thin layers that decrease in grain size from bottom to top. Identical layers are common on the sea floor.

Sediments tend to move over the edges of the continental shelves and widen the shelves by blanketing the outer slopes. But the material, more water than mud, is delicately poised, and prone to cascade down the slope and far across the sea floor when triggered by any one of a number of such causes as earthquakes and tsunamis. Recent drilling in the oceanic deeps has brought up from many sites sediments that must have had their origins in these far- and fast-travelling suspensions.

We used to think that sands should all be next to the shores and muds should be in deeper water beyond them, but submarine sampling and mapping show patterns more complex than this. Deep currents capable of moving sediment are more numerous than anticipated by early students of the oceans.

In the ocean depths, many miles from shore, the rate of accumulation of debris is impressively slow—a small fraction of an inch in thousands of years. In areas where organisms are abundant in the upper waters, a small percentage of shells survives the long trip down to the bottom after the organism has died. Shells are subjected to long exposure to the cold bottom waters, slowly accumulating to make the deep sea oozes. Some oozes are made up dominantly of calcium carbonate shells from forams and coccoliths, others of the silica-containing shells of diatoms and radiolarians. Beneath the areas of low surface production of these organisms, the red muds are found, a roughly 50–50 mixture of clay particles brought in by streams and of dust blown far across the sea from land.

One of the scientific surprises of the last few years has been the discovery of the importance of wind-blown material in the deep sea deposits. Joseph M. Prospero, a scientist from the Marine Institute at Miami, Florida, working on the Island of Barbados at the eastern edge of the Caribbean, set up fine-meshed cloth screens coated with a sticky substance to catch dust. The trade winds blow at twenty miles an hour uninterruptedly across the Atlantic to Barbados, so that every day four to six hundred miles of wind passed through the screens, which trapped suspended solid particles. After a day or two the screens turned brown and yielded abundant muds when they were rinsed. Examination of the minerals in the muds showed that they were the same as those blown about in dust storms on the Sahara desert. Moreover, the dust catch was greater a few days after a big storm on the Sahara—just the length of time it would be expected to take to reach Barbados, as calculated from the wind direction and velocity.

The deep sea deposits west of Australia for a thousand miles are dominated by clay minerals blown from the deserts of Australia. The northwest Pacific has great plumes of dust on its floor, blown eastward from the Asian continent. Despite this, the total contribution of air-borne dust to the oceans is miniscule compared to the contribution of material from rivers and only the sediments far out to sea have a substantial air-borne component. This is because the land-derived material settles out close to shore.

The slow rate of accumulation of settling particles far from land is also indirectly responsible for the formation of manganese nodules on the ocean floor. In many areas the bottom is dotted with irregular knobby chunks of manganese oxide, in some places so thick that they make a continuous pavement. The nodules are a potential source of manganese and several other metals such as cobalt and nickel. The total on all the ocean floors aggregates hundreds of billions of tons of manganese oxide. Although some nodules occur at shallow depths—a few hundred feet—most of them lie two miles or more below the ocean surface and constitute a difficult and expensive recovery problem. No one is quite sure just how they form, except that they have grown where they are found. The manganese may come from sea water, from the sediments beneath, or from both sources.

Deep sea sediments, as a whole, are relatively rich in quite a few elements, in contrast to the occurrence of these elements in the average land rocks. In addition to manganese, cobalt, and nickel, the sediments are enriched in lead, molybdenum, and barium. Some day they may be valuable sources of these metals, though this looks far off at the moment. The problems of recovery are similar to those of recovering elements from sea water. There is a lot of lead and nickel and cobalt on the floor of the ocean, but it is spread out over a huge area that is not easily reached.

An intriguing discovery was recently made in the Red Sea. During routine measurements of depth, temperature, and salinity, oceanographers were astounded to find the water temperature climbing higher with greater depth, rising to over 100° F near the bottom. As the temperature rose, the water became many times more salty than ordinary ocean water. Further investigation showed a pocket in the sea floor filled with dense, hot, salty water, apparently heated by volcanic activity far below. The sediments encountered at the bottom also were unique, consisting of iron oxides high in zinc and other heavy metals. This sediment discovery is exciting, for it is a new kind of deposit and has definite economic possibilities. It is also obviously a political hot potato—who is entitled to benefit from exploitation of the deposit?

Oil is an unlikely component of the slowly accumulated deep sea deposits, but it has already been exploited avidly on the shallow continental shelves. The occurrences on the shelves are like the occurrences on land; they are oil and gas that have accumulated in old rocks and have nothing to do with sediments forming in the present-day ocean. Because the deposits have to be exploited by drilling through the sea into the rocks below, any oil leaks during drilling and recovery cannot be run into a pit in the ground and recaptured as they are on land; instead they spread destruction across the sea surface. There has been enough publicity about the leaks like those at Santa Barbara, California, with the resultant fouling of beaches, killing of birds and fish, and other unfortunate side effects, to bypass the details of the story here. The ownership of oil on the continental shelves of the United States has caused many fascinating legal problems. Does it belong to the adjacent state or to the federal government? How far out are the national boundaries? Can one drill for oil offshore of another country? Should there be a three-mile limit, a twelve-mile limit, or a depth limit to the control of the seas? And with the increasing incidence of spills and leaks, designation of responsibility for the damage is becoming increasingly important. Huge gas deposits have been found in the North Sea between Britain and Scandinavia. Their exploitation has been worked out quite amicably by the adjacent countries. But it is evident, as exploration of the oceans continues, that a lot of effort will go into deciding ownership of the sea and its floor.

Diamonds have been mined from the sea floor; like oil, their occurrence is an extension of their land occurrences and not a truly marine phenomenon. Diamonds are hard, heavy, and durable, so they tend to remain in stream gravels while many other minerals dissolve or are abraded away. Diamonds have been found in stream beds of South Africa and they have been taken from beach gravels, where waves have distributed the stream loads along the shore. Gravels containing diamonds also are found in shallow water beyond the beaches and have been mined by dredges. Tin also occurs in gravel, and it too has been mined in the shallow seas.

Very broadly it looks as though the continental shelves, being really part of the continents that are slightly awash, have the same mineral potential as similar areas on dry land. If so, their economic importance, now being exploited rapidly, should amount to about 15 percent of that of the land. Except for the Red Sea iron and zinc occurrence, which has not yet been exploited, and some oil and gas possibilities in deep waters of the Gulf of Mexico, the mineral potential of the deep ocean bottoms is not highly encouraging. Also, there is a considerable technologic barrier to mining beneath a mile or two of sea water. Any deep oil and gas occurrences will probably be put into production before solid materials are mined. It is much easier to bring liquids or gases up from the depths than it is to pull up solids. Liquids and gases can be allowed to more or less float their way up.

The real floors of the deep seas, on which the sediments are a thin carpet, are mostly basalt lava, none of which seems to be older than two hundred million years. Yet fossil marine organisms in the rocks of the continents are evidence of seas billions of years old. It must be concluded that the floors of the ocean basins are made of lavas that have come from the depths of the earth in the comparatively recent geologic past (two hundred million years is young compared to the earth age of

4.5 billion). The processes going on in the ocean basins today are almost beyond imagining. The great mid-Atlantic ridge is a place where the earth's crust, stressed from deep inside the earth, is splitting and widening; the crack, as it opens, is filled by lava rising from the interior of the earth. The area of Atlantic Ocean floor increases, continents drift apart. This hypothesis is well enough documented that we can say with certainty that Africa and South America were once close neighbors, as were North America and Europe. North and South America have separated from Africa and Europe, "floating" plate-like on the "crust" below. If one likes to make really long-term investments, he might buy an air ticket from New York to London and hold it, because next year the distance from New York to London will be an inch greater, while in sixty thousand years the ticket holder will get an extra mile of flying for nothing! The evidence suggests that all the continents were once in a single mass, and that there was just one great ocean. During splitting apart and moving of the continents to their present positions, they have moved differentially relative to the poles and the equator. Rocks of Antarctica contain fossils of tropical animals, recording the travels of that ice-bound land. The Atlantic basin could not have been created without obliterating an equal amount of the Pacific basin, that is if the sizes of the continents have not changed. Also, one would expect that what remains of the Pacific basin should be older than the Atlantic, but no such thing. The same kind of cracking and filling by lavas is taking place in the Pacific. But the ocean floor of the Pacific, instead of widening and pushing the continents with it, is plunging beneath the continents. The great "ring of fire" of volcanoes and earthquakes that surrounds the Pacific basin is indirect testimony to the conflict between ocean floors and continents. Very grossly, all the ocean basins have rifts that continue to widen and fill with lava that "freezes" to make new sea floor. The old floor must go somewhere and, like dust under the rug, a good bit of it is being pushed beneath the circum-Pacific continents.

136

The deep sea sediments, accumulating ever so slowly on their moving carpet of frozen lava crust, average only about a half-mile thick. Most of them are still soft and wet, about 40 to 80 percent sea water, and still too soupy to be called rocks. The water trapped in them, in most places, is like present-day sea water, showing that the oceans have not changed their composition for many tens of millions of years.

Exploration of the sea has been full of surprises. An analogy to the discovery of the Americas is probably a good one. The depths of the oceans will not yield piles of gold and diamonds; and get-rich-quick schemes, like those of Sir Walter Raleigh, who brought a shipload of fool's gold back to England, will fail. It will take time to realize the potential of what has been discovered in the seas. But the variety of riches that has emerged, the mere fact that exploration of the sea has already completely changed older ideas about the origin of the ocean basins, suggests that its contributions in the future will be like those of the New World. George III could not conceive that the Colonies would become a world power or that their strength would come from mines and cornfields unknown to the Thirteen States. The true potential of a new domain does not emerge immediately. It is judged by the accomplishments of the old ones and is inconceivable before it is realized.

~~~~~~

CHAPTER EIGHT

The Shallow Ocean

SHORTLY AFTER World War II the United States found itself in possession of a large volume of radioactive wastes created by the development of atomic bombs. These wastes were of many different kinds and posed both short-term and long-term hazards to living organisms. One of the first suggestions for their disposal was to put them into the ocean deeps, and the advice of oceanographers concerning the safety of deep sea disposal was solicited. Then, for the first time, the capacity of the world oceans as a dumping ground was assessed. Scientists still cannot answer many of the questions that were raised, but because of the pressures put on them, they have been struggling valiantly to provide them. Characteristically, the problem is not answerable by a yes or a no. To say whether a given material can safely be put into the sea, it is necessary to know the nature and amount of the material, how much it must be diluted to become harmless, where it will be put and how, whether it might be concentrated by organisms, and a thousand other pieces of information. To give unequivocal answers to all the questions, the oceans must be understood in great detail. Information required for answers includes knowledge of the circulation of the oceans, the rate at which the waters mix, the interplay between atmosphere and sea, chemical communication between the floor of the sea and the water above it, the effects of waves and tides and related water movements that play such important roles in near shore areas.

Deep sea disposal, as it turns out, is a rather special situation, and has been considered chiefly with respect to some of the most hazardous substances. The cost of putting something into the ocean deeps is so great that this is usually considered only for extremely toxic materials, for substances so dangerous that safe land storage cannot be guaranteed. In general it has been better to keep toxic materials on land than to put them into the mobile deep ocean. At any rate, the conclusion has been drawn that the ocean deeps are not the place to get rid of really bad wastes, and they have not been used. Instead, in the past twenty-five years, attention has been centered on coastal areas, and the problem of contamination of coastal waters by radioactive substances has broadened to a consideration of pollution of all kinds, including that from sewage, oil, and a variety of chemicals.

The fate of a pollutant introduced along a coast involves waves, tides, and all the other phenomena that influence the interaction of sea and shore. This chapter is devoted to exploring some of these shallow sea processes.

It is the interplay of wind and water that creates the waves that give the oceans such variety. When the wind blows, water cannot resist the drag and remain flat, it must hump and hollow as it tries to follow the wind. That this should be is strange; wind should quell the waves, not make them. In the open sea the waves are *of* the water and yet not water itself; they move but it does not. The water merely circles as the waves go by, rising and falling in its place.

It is the wave *shape* that travels, not the water, just as a wrinkle in a sheet moves across the bed as the sheet is smoothed out. When the wind begins to blow across a calm sea and finds some chance-created tiny hill or hollow, the air currents are disturbed. They push and pull on the hill and make it a little bigger for a moment; it sinks down again, the air rushes on and pushes up another little hill downwind. So each tiny hill becomes the cause of yet another, until the whole sea surface is rippled. Each ripple disturbs the wind yet more. The ripples grow higher and more widely spaced until they are waves. They grow in height and spacing as the wind continues to blow. If the wind is steady and holds its direction for many hours, the waves and wind finally become adjusted to each other. The height and speed and spacing of waves stay constant until the wind changes.

The harder the wind blows and the longer it blows and the greater the distance across which it blows, the bigger the waves that are formed. Happily there is always a limit on the size of the waves, no matter how long the wind endures. If the heights and speeds and spacings are measured, regular and simple relations of these properties are found. The period of a wave—the time it takes one crest to occupy the position of the preceding one—increases regularly with the speed of the waves; the spacing between crests increases with the square of the speed. If a ship anchored to ride out a storm and the Captain diverted himself by observing the waves, he might find that in the early stages of the storm the wave crests would move past him at 20 miles an hour at 5 second intervals, and that the spacing between crests would be 150 feet. When the waves speeded up to 40 miles an hour, the interval would have increased to 10 seconds, and the spacing to 600 feet. Doubling of wave speed doubles the period but increases spacing fourfold.

There are fierce arguments over the greatest heights reached by waves in the open sea. Some of the difficulties in measuring accurately are obvious. To make a measurement from a ship that is being battered by a fierce storm is almost impossible;

the deck tilts, waves crash over the bow, the wind is too strong to stand against. Single waves as high as one hundred feet have been reported by reliable observers. Successive waves of forty to fifty feet in height are probably the largest formed in all but the exceptional storm.

Two sets of waves traveling in different directions can cross each other without either set being destroyed. When a wave crest of one set meets the crest of the other, the water heaps up higher; when two troughs come together, the valley formed is lower than that of either single wave; where a trough and a crest coincide they cancel each other. Interfering waves, as they are called, can produce water mountains higher than crests of either set alone.

Once waves have been generated by a storm they can outrun the storm, moving beyond the place where their generating wind is blowing. Technically, swells are exactly such waves; they account for the continuous surf that pounds in on many beaches, irrespective of the local wind direction. Once formed, waves propagate themselves efficiently indeed. Swells created in the South Atlantic can be dangerous to fishing boats as far away as Newfoundland; storm waves generated near Hawaii draw thousands of spectators on the coast of southern California, twenty-four hundred miles away, where the swells cause spectacular breakers.

Surfers are keenly aware of wave interference. When two sets of swells are moving onto a beach from nearly the same direction, and with almost the same wave spacing, there may be several minutes during which the crests of one set of waves nearly coincide with the troughs of the second set, cancelling out both and creating a calm sea. Then, after a while, the crests of both sets coincide for a few waves and the waiting surfers ride these big ones in. This phenomenon of "beats," as it is called, is especially pronounced on the north coast of the island of Oahu, Hawaii. There the normal winter surf is spectacular, but when two sets of waves come in and their crests coincide, a truly huge comber smashes in, sometimes completely sweeping the beach of swimmers, picnic gear, and anything within the limits of its uprush.

The circular motion of the water that takes place as a wave goes by dies out rapidly downward. At a depth about equal to half the spacing between waves, a diver is hardly aware of waves that would toss him about like a cork at the surface. Submarines can move placidly through a hurricane by submerging a few hundred feet.

The bigger swells may have spacings of about fourteen hundred feet and travel sixty miles an hour. The record is a spacing of twenty-six hundred feet and a speed of eighty miles an hour. Many boatmen have had the eerie experience of riding in the trough of a good-sized wave that has the same speed and direction as the boat, so that the crest looms continuously and threateningly over the stern, giving the sensation of being propelled by the wave.

Thanks to waves, the upper five or six hundred feet of the oceans are well mixed and have abundant oxygen. As waves build up in a freshening wind they become steep and the tops are curled and blown over by the wind to make the familiar whitecaps. The white color of the capping water comes from the millions of bubbles formed when the curling crest traps pockets of air and churns them into the water. About a third of the entire sea surface is covered with white foam at any one time. The air of the lower atmosphere and anything it contains is effectively mixed into the upper layers of the ocean in this process.

This frothing action of the waves causes many important chemical reactions. The bubble surfaces act as collectors for various substances in the sea water and may be responsible for changing the chemical form of some of the dissolved organic material, transforming it from inedible substances into food for various organisms. The churning of the sea and the breaking of the bubbles as they surface throw a fine mist of sea water into the air. This water evaporates, leaving its salts behind as tiny solid particles. Some of these are swirled high up into the atmosphere and eventually settle out or are carried down in rain. This "dry" and "wet" fallout has a strong influence on the environment of coastal regions. Some plants are killed by the salts in rain, others can survive. The seashore flowers, shrubs, and trees are those that can tolerate the relatively high amounts of salt contributed by the sea. The salts are also remarkably effective in causing corrosion of many materials. Coastal dwellers know that automobiles deteriorate in a seemingly magical way and that houses have to be painted more frequently than do those a few miles inland.

One of the mysteries of these air-borne materials from the ocean is that their composition is not exactly the same as that of salts in sea water. Somehow this bubbling and frothing that goes on during their production manages to select some salts in preference to others. The composition of rain in Hawaii, although much like very dilute sea water, has a higher proportion of calcium and a lower proportion of sodium, than sea salt, and differs in other constituents as well.

A few miles inland from most coasts, the concentration of dissolved sea salts in rain drops is but a few parts per million, but the total amount of material from the sea that is dumped on the continents as rain is huge. About half of the sodium and chloride that rivers bring to the sea comes from the sea and not from leaching of the land. A substantial fraction of the calcium and sulfate have the same origin.

In a drifting boat on the open sea when it is calm but with a strong swell running from some distant storm, the relation between the movement of the waves and that of the water is most apparent. The boat bobs endlessly as the swells sweep past with racehorse speed. If the waves are being driven by the wind the great bulk of water still circles as the waves go by, but a slow surface current, moving in the direction of the wind, develops. Otherwise there is no important mass movement of water related to the waves. As waves move into shallow water the situation changes. The waves are no longer free to die out naturally for the bottom restrains the circling motion of the water. In response, the bottoms of the waves are slowed, the crests are forced to rise and steepen and they topple forward to create breakers. On a gently shelving shore, a wave may break continuously for hundreds of yards as it moves in. The energy of the deep water wave, which extends downward a half wave length or so, is progressively eliminated by the shallowing sea floor and makes the wave keep peaking and breaking. The wave is constricted from below and forced to rise and topple, like toothpaste squeezed from a tube.

On long stretches of beach, waves always seem to move directly toward the shore. If the beach has a strong crescent shape, the crest of each wave as it moves in and breaks has a crescent shape too, parallel to the shore line. Yet the waves approaching the shore from the open ocean come from one direction on one day and from a different one the next. It is the bottom slope that shapes the wave crests. When deep water waves move straight in to a straight shore with a uniform bottom slope they keep their straight crestlines, but a view from above shows that the breaking crests get closer and closer together as the waves are slowed on their course to the beach. If there is a deep channel extending outward from the center of a straight beach, the waves are less impeded by the bottom there and their crests bend toward the shore above the channel. Breakers form nearer the shore in the channel area. When waves come into a uniformly straight beach from an angle the wave crests tend to bend around to approach the shore nearly straight on because the part of the crest in deeper water moves faster than the part that has "felt" the bottom and the wave thus swings around like a column of marching men doing a "right wheel."

From aerial photographs of waves moving toward a shore it is possible to map water depths, the shoals, and the channels. In World War II the best sites for amphibious landings on unmapped shores were chosen by studying wave patterns in aerial photographs.

When a breaking wave finally comes up on the beach its

"bottom" has been so retarded relative to its "top" that it topples over to make a sheet of water that dashes up on the beach and drains back into the sea. Only in the breaker zone does the water move with the wave and even there the movement is in and out with each wave. The circular movement of the water of open ocean waves is changed into a horizontal ebb and flow. On some beaches, when the breakers come at short intervals, the water returning from the uprush of the waves becomes channeled in its return flow and rip currents result. For the unprepared they are dangerous for they can carry a swimmer out through the breaker zone into deep water faster than he can swim against them. Surfers who know the local water often use rip currents as highways to save themselves the hard work of paddling out through the breakers for their next ride.

The combination of tide and breaking waves gives exposed sand shores a good message during the tidal cycle. At the first break of a wave the crest cascades downwards like a waterfall, stirring up the bottom and putting sand into suspension in the water. When the tide is low the breaking waves dig deep into the bottom, moving great masses of sand outward. At high tide the material is returned and filled in. No wonder that beaches are sandy; any fine muddy material drifts out into deep water because of the constant stirring, while heavier sand settles out on the beach and in shallow water. The greater activity in the wide breaker zone of exposed coastal beaches is a wonderful protection against pollution. Waste materials such as sewage and harmful bacteria are quickly oxidized and destroyed by the swirling bubbly water. Even sharp broken glass is quickly ground and smoothed into harmless rounded shapes by the abrasive sand.

On rocky coasts the water is often deep right up to the shore and the open ocean swells dash directly against the cliffs. Where the rocks are hard the waves do surprisingly little damage, even though they may crash against the rocks and throw spray dozens of feet into the air. In Nova Scotia, hard granite, scoured and smoothed by glaciers thousands of years ago, slopes smoothly into the sea without even a nick cut at the waterline by wave action. But if the coast is rugged and the rocks are soft or poorly cemented, waves can loosen the mineral grains and undermine the shore so that great blocks of rock eventually tumble down to accumulate at water level, until they in turn are ground to bits. Waves gradually cut a rocky platform about at low tide level. Sometimes little spires or islands of harder rock stick up above this wave-cut bench. Here and there along such coasts a zone of softer or more broken rock yields to the waves and small coves with pocket beaches are cut back into the cliffs. No matter how soft the rock, the coves cannot cut inland indefinitely for the waves are curved and "stretched" as they enter the cove and their energy becomes dissipated. High shores, therefore, with rocks of variable hardness make the most spectacular coasts. High cliffs, rocky offshore islands, pocket coves with curving beaches—these are all the result of the action of waves. If the coast is high, but soft and uniform in its materials, a cliff and a beach develop, but the cliff is straight and the beach is narrow and continuous. In places one must travel many miles to find a way down to the beach. On such shores there is no protection for a boat; if one does get into trouble during a storm it must run many miles for shelter. If driven ashore it will be pounded to pieces by the storm waves dashing across the beach to the base of the cliffs. Lake shores can have the same features as ocean shores. The treacherous west shore of Lake Michigan is a good example. The eighty miles of straight shore between Chicago and Milwaukee is almost unbroken cliffs; boats stay well offshore in stormy weather.

Waves seldom come in exactly parallel to the shore. As a result they slosh up on the beach at a slight angle, then drain back straight down the beach. This small deviation from a straight up-and-down-the-beach movement causes a current to develop along the shore, the strength of which depends on the size of the waves and the obliqueness of their initial approach. The effect of the combination of wave slosh and current is to transport materials of the beach and of the shallow water zone along the beach. On beaches that receive waves equally from all directions, the material is just shifted back and forth, but if most big waves come in from a particular direction, the net movement of sand each year is also in one direction. Such a situation is the common one and leads to some interesting problems, especially for owners of beach property. If I have a home on the shore, my sand today is my neighbor's sand tomorrow and was my other neighbor's sand yesterday. All is well if I receive as much as I lose, but if my up-current neighbor wants to make his beach wider he may decide to build a pier out into the water at the boundary of our properties to hold his sand in place and to catch whatever sand is moving. If he does this and the supply to my beach is cut off I may build a pier to keep what sand I have, thereby removing the sand supply of my down-current neighbor. No wonder the business of beach protection and preservation has caused many legal battles and has become a matter for State and Federal policy.

The sand for many beaches is not supplied by direct erosion of the shore by waves but is brought to the sea by rivers and distributed by waves and currents. In many cases dams have been built on some river far upstream from the coast, perhaps to generate electric power or provide irrigation water, and after a few years the coastal inhabitants have found their beaches shrinking and disappearing as the dam prevents the sand supply from traveling downstream. The legal complications are formidable, especially if the dam is in one State and the beach in another.

Tides enhance the action of waves and create their own currents. As the earth turns daily beneath the sun and moon, its water and its land as well are attracted by these celestial bodies to make the tides. When both sun and moon are on the same side of the earth, the tides are highest (spring tides), and when they are on opposite sides, lowest (neap tides). The tidal bulge is almost imperceptible in the open ocean, but because its crest is racing westward a thousand miles an hour at the equator, it piles up on the coasts of the continents and is accentuated where there are deeply indented bays or inlets. At first consideration it appears that there should be but one tide per day, but it turns out that when the attraction of the moon (which is the more important force) raises the waters on the side of the earth nearest to it there is a weakening of the earth's gravity on the far side that lets the ocean surface rise there as well. The normal situation is two high and two low tides a day, but because of channels, interfering currents, and other influences, some places, like Manila in the Philippines, have only one tide a day, while at San Francisco there are two highs and two lows, but the two highs have strikingly different heights.

Because the moon rises later each day, the times of high and low tide change daily by about an hour. On an island like Hawaii, in the open ocean and without any places where the tide can be funneled and concentrated, the ocean level changes only a foot or two from high to low tide. In relatively small bodies of water like the Great Lakes, the tide is imperceptible, and even in the Mediterranean Sea it can hardly be noticed. But on the French coast of the English Channel, where the tidal bulge is concentrated, the water level may change as much as forty feet between high and low tide. In Nova Scotia's Bay of Fundy, tides reach sixty feet and a tidal "bore" ten feet high runs up the rivers that feed into the Bay.

The tides have had an important role in history. Perhaps because of his experiences with the miniscule Mediterranean tides, Caesar forgot to allow for the change in water levels when he landed on the shores of Britain. When the tide came up some twenty feet or more his galleys were wrecked. The Allied invasion of France in June 1944 almost failed because of the difficulties of landing on and supplying a shore with thirty foot tides; on the other hand, that section of shore was chosen partly because the tides were so high that the Nazis had it more lightly defended than elsewhere.

Ocean shores owe much of their attractiveness to the tides. At high tide only a tiny strip of beach may be visible; at low tide it may be hundreds of feet wide. Swimming conditions change according to the stage of the tide; organisms are stranded on the beach as the tide falls. On rocky shores pools of sea water are left behind at low tide; they are inhabited up to the low tide level by organisms that must be submerged to live. Between low and high tide levels the rocks are occupied by parallel bands of marine creatures, with those most resistant to drying and exposure to air at the very top. Tide pools make wonderful aquaria where marine creatures can easily be observed.

When the great Alaska Earthquake occurred in 1964, some parts of the coastline were raised or lowered as much as forty feet. The most reliable indicators of original high tide level are barnacles. These tough little shellfish, although they cannot live in fresh water or in air, can spend most of their time out of water as long as they are wet from time to time. Even now their shells can still be seen attached firmly to the rocks along the uplifted coastal areas.

Although the general effect of the tides produces a flushing and renewing action, there are many coastal areas in which the net result of flooding and ebbing is to deposit mud in shallow depths, developing tidal flats. Where tidal flats are very wide they can be treacherous to man. It is easy to get caught on foot too far from shore to outrace the rising tide. Even for a boat, the combination of shallow water, currents, and breaking waves may be disastrous. Off the west coast of Korea, which has a large tidal range, the flats are many miles wide and make navigation difficult.

Tidal flats are important sources of sulfur gases in the atmosphere. When the tide is low, bacteria are hard at work in the expanse of wet mud, reducing the sulfate in sea water to hydrogen sulfide, which escapes to the air, producing a strong odor locally, and is oxidized by the air to various other sulfur gases. Eventually these gases are dissolved in raindrops and become a source of sulfate in river water. Our lack of knowledge of the importance of the process has been largely responsible for difficulties in assessing the amount of sulfur gases added to the atmosphere by the burning of fossil fuels, but anyone who lives near an extensive, muddy, tidal flat is willing to swear that the production of hydrogen sulfide at low tide is far from negligible.

Many attempts have been made to harness the power of tides, but so far with only moderate success, even though early grain mills along some coasts were operated by tidal power. The difficulties are related mostly to the rise and fall of the tides. Most tidal power plants are based on letting a basin fill at high tide, impounding the water, then discharging the water as the tide falls. On its way in and out of the basin the water runs a turbine generator. The height down which the water flows changes with the level of the sea outside so that the power that can be generated goes from zero to a maximum and back to zero again. One of the prime requirements of electrical power in modern times is continuous generation and the ability to deliver on demand. The changing time of high tide, as well as the variations in the level of high tide, have caused many difficulties. The most economical installations would be those with a large tidal range and a tremendous water storage capacity and electrical output. This requires a huge capital investment. Unfortunately, the best geographic locations have not coincided with large local power demand and power that has to be transported is expensive. The Bay of Fundy is too far from major power consumers to be an important power source. An estimate by M.K. Hubbert of the U.S. Geological Survey of the obtainable annual tidal energy is one hundred billion kilowatt hours—about 4 percent of the 1964 world electrical power production and an even less important percentage of today's output.

On the other hand, hydroelectric power is "clean" power. The cost differential between tidal power and other kinds of power is diminishing with the current trend to force power producers to include in their prices the cost of protecting the environment from their waste products.

"Tidal waves" are quite unrelated to the daily tides. The Hawaiians call tidal waves *tsunamis*, and the use of the word is spreading. To Hawaiians the distinction is clear. Because the islands are far out in the Pacific, the tidal range is small. There-fore, when irregularly, over intervals of years, the ocean rises up and dashes a great wall of water against the shore, it is clear that the event is not just tidal action and it deserves a special name.

Tsunamis are generated when the sea is shocked by some abrupt and violent event, like an earthquake, a volcanic explo-sion, or an underwater nuclear explosion. In an earthquake the sea floor is displaced along a crack and the whole water column above is disturbed, sending out a short train of waves. The height of the resulting wave, or bulge, is only a few inches in the open sea, but the wave travels at speeds of hundreds of miles an hour. When it impinges on a coastline it builds up into a great wall as the water shallows. In the 1964 Alaska Earthquake, the sea floor was displaced as much as forty feet. Only a series of near-miraculous coincidences saved thousands of lives. The earthquake struck on Good Friday morning when schools and factories were closed, dock workers were at home, and the waterfront was nearly deserted. The tsunami created by the earthquake topped the trees of Alaska's coastal forest. As it moved southward along the North American coast, it de-stroyed shore facilities such as wharves and warehouses, and smashed boats, throwing the debris up on land. In Alaska, the salmon-canning factories located at the water's edge were par-ticularly vulnerable. Within a few hours the wave had done substantial damage as far south as Mexico.

Since the series of great waves that struck Hawaii in 1946, an International Tsunami Warning System has been in opera-tion. Because the waves travel so fast, they reach coasts thou-sands of miles distant within a few hours after their inception. The System monitors all information pertinent to tsunami for-mation and movement and broadcasts warnings to any place in the world that might be affected. In Hawaii, sirens have been installed along some of the broad low coastal areas that could be inundated by a tsunami.

It is fortunate that the great tsunamis are relatively infre-

151

quent. The low exposed coastal areas of the world are becoming increasingly populated as more and more people seek the advantages of work and recreation at the water's edge. Even though the wave strikes briefly, it wipes the shore clean within its reaches. A tidal wave in 1929, caused by an earthquake at Georges Banks south of Newfoundland, crippled the south shore of Newfoundland for many years. The wave entered all the harbors of the fishing villages and smashed the ships, then continued up the rivers that flowed into the harbors, wiping out the bridges that linked roads along the deeply indented shoreline. The Newfoundlanders found themselves unable to fish and thus unable to afford to replace the bridges. For many years the villages remained isolated and only subsistence fishing could be done from small boats.

Intense tropical storms—hurricanes in the Atlantic and typhoons in the Pacific—have destructive effects like those of tsunamis as they approach a shore. Not only do they generate high waves, they accentuate the normal tidal range as the winds literally pile the water up ahead of them by generating strong surface water currents. This permits the great waves to pound the shore well above the ordinary high tide level. At Cape Cod, the general water level was raised several feet during one hurricane. The main building of the Woods Hole Oceanographic Institution at Falmouth became an island. Waves dashed across beaches that normally separated the ocean from fresh water lagoons, and the salt water killed fresh water plants and animals. At Miami, Florida, where many homes are built on flat lands a few feet above high tide, large areas must be completely evacuated when a hurricane approaches. Even in areas protected from waves, flooding may occur.

In addition to tides, tsunamis, and wind-generated waves, there is another phenomenon called a *seiche*. It occurs in lakes or harbors or other relatively restricted water bodies. It is more a giant "slosh" than what we ordinarily think of as a wave. The water in a basin oscillates back and forth like coffee in a carried cup or like bathwater as one plumps into the tub. The seiche is often triggered by a sudden local change in atmospheric pressure, such as accompanies a thunderstorm. Just as the coffee can be set in motion with an oscillation period of a fraction of a second by blowing sharply into the cup, the atmospheric pressure change causes a longer period vibration of the water in a lake basin. The period of the seiche increases with the size of the basin. On Lake Michigan, for example, it takes about twenty minutes for the water to rise and fall. A seiche is remarkable to see for the first time, especially without prior knowledge of its behavior. One of the major seiches on Lake Michigan happened one summer day. The lake level began to drop slowly but steadily for many minutes, for a total of several feet. On many bathing beaches this meant that the water line moved out tens of feet. Then the water returned more quickly than it had retreated, rising to several feet above the original level. Many curious bathers who had quite humanly followed the retreating water became engulfed, but the rising water didn't have enough speed to endanger most of them, although several persons who had been fishing from a long, low pier were washed off and drowned. After the first seiche, the water continued to oscillate for hours with a constant period.

Seiches are also caused by tidal currents entering partly enclosed harbors. There are many small basinal harbors along the coast of Japan that are particularly susceptible to seiches. Special pilots with experience of local conditions are employed to dock ships in these harbors. In an enclosed harbor, a mile or so across, the water sloshes back and forth regularly every few minutes; usually the effect is not strong enough to be noticeable, but variations in day to day conditions can at times cause changes in level of several feet. A tsunami in the Pacific may set all the harbors along the shore into a seiching motion that can last for days.

Waves, tides, tsunamis, and seiches have their major influence on the ocean shores. They are responsible for the mixing of the growing efflux of waste from land with the water of the ocean. For most substances, the zone of transition from their delivery

153

at the shore by streams to their dissipation in the open ocean is the critical area of concern today. Although there are some pollutants that are worrisome even if mixed throughout the body of the oceans, there are many more that would be harmless if it could be assured that they would be dispersed adequately. The simplest example is that of heat disposal. Discharge of hot waste water into a restricted volume of sea water can cause temperature increase and important effects on marine life, but the same discharge into major deep cold currents would have an unmeasurable temperature effect.

Where the coastal sea is shallow, the nearshore zone may be tens of miles wide, as on the Atlantic coast of the United States. Where deep water comes close to the shore, and where major ocean currents can impinge on the coastline, most land-derived influences are quickly dissipated to undetectable levels by mixing. The transition zone is narrow—perhaps a mile or so wide. It takes time for wastes to move out of the nearshore zone as they are carried erratically back and forth along the shore. There is a zone of mixing where sea water moves in and land-produced products move out. The mixing times are long enough, and the amount of pollutants discharged are already great enough to raise the general level of contamination in some discharge areas to undesirable levels. On the Atlantic coast of the United States, an average molecule of water from the Savannah or Delaware Rivers takes about two years to work its way out to the open sea. One of the most difficult aspects of discharge to the nearshore is that an amount of waste that would be dispersed harmlessly on nine hundred and ninety-nine days out of a thousand may attain a dangerous temporary local concentration as the result of a fortuitous combination of wind, tide, and swell. On the happier side, it is fortunate that the nearshore zone is the place where the waves expend their energies. Exposed coasts can "digest" many times as much waste as can protected ones. Most of the problems of marine pollution cropped up first in harbors, lagoons, and estuaries, where protection from the waves attracts ships and supporting shore facilities. Without rigorous controls, all kinds of wastes tend to be dumped into exactly the places where their destruction and dispersal are slowest.

Much attention is focused today on estuaries—the lower courses of rivers where fresh and salt water mix. Many estuaries of the Atlantic seaboard of the United States are the sunken mouths of rivers; the Hudson, the Delaware, and the Potomac are representative. Thousands of years ago the land was much higher, relative to sea level. The rivers cut broad valleys in their lower courses and when the sea rose the valleys were "drowned" for many miles upstream. In the drowned part of the valleys a complicated system developed involving the fresh water entering from the river and the sea water moving in and out under the influence of the tides. In estuaries, the change from fresh to salt water is spread over a distance of many miles. This transition is expressed in a zoning of organisms, with fresh water forms giving way to those of greater and greater salt tolerance as the open sea is approached.

If a dense chemical waste is piped into an estuary, it will sink into the salt water layer and will be carried up the estuary, gradually mixing with fresh water as it goes. Because of tidal currents, it will work its way up in a series of oscillations, but eventually, following the overall "up-bottom down-top" circulation, will get into the surface water and be carried out to sea. There is a virtual guarantee that dense wastes will be widely distributed throughout the whole system.

If the wastes are dilute and less dense than salt water, they will move out of the estuary more quickly, following the surface flow. Even so, they will fan out as they mix horizontally, and some of the material will get into the deeper water of the estuary from the mixing at the fresh water-salt water interface.

All this assumes that the waste material stays in solution. But it may interact with the sea salts and be precipitated to sink out as a solid, or it may be abstracted by organisms. (It has been calculated that all the water in Chesapeake Bay passes through oysters every day as they pump it through their systems to obtain food. The opportunity for removal of waste material from the water or for concentration of a substance in the oyster's tissues, is obvious.)

Because of the mixing and dispersal of pollutants by these various processes, estuaries are difficult systems to control. They are not like rivers—with a relatively small flow in one direction, and therefore fairly manageable. Their erratic flow and large volume, plus the long residence time of pollutants, makes them difficult to clean up. They become reservoirs for undesirable substances.

Even on an open shore exposed to waves and tides, where mixing of a local discharge of wastes is much faster than in an estuary, there are many factors that make it difficult to be confident that dispersion of pollutants to innocuous levels will always take place.

The current pattern on a shallow shore is variable. The direction of the longshore current in the breaker zone depends upon the direction from which swells approach the beach. The direction may change from day to day. Outside the breaker zone, surface drift currents caused by the local wind may move in the same direction as the longshore current, or opposite to it. Superimposed on these currents are those related to the rise and fall of the tide, which move back and forth nearly parallel to most shores. As a result of all these currents moving in various directions, waste waters added from a pipe or small stream at the shore follow a sinuous course as they flow out and mix with the sea—in one direction along the shore in the breaker zone, perhaps in the reverse direction when they get outside, then doubled back on themselves as the tide reverses. Typically, a low general level of concentration of the wastes builds up over a considerable area in the vicinity of the source, with a plume of wastes of higher concentration reflecting transient addition. How long it takes a given amount of waste to make its way out to the open sea depends on many factors. A strong offshore breeze will blow surface water outward; it is replaced by deeper, colder water moving in from underneath and creating a major circulation that speeds mixing and the transport of waste. On the other hand, when such a wind stops, the displaced surface water tends to move back toward the shore and this is the time when wastes tend to be thrown up on the beaches. Such an event, even if it occurs only once in several years, can create a serious problem.

Today there are encouraging advances in the ability to predict in detail what will happen to waste discharged into estuaries or along open coasts. With high speed computers it is possible to build mathematical models to simulate natural conditions. The computer can use numerical values for water temperatures, salinities, current speeds, tidal conditions, amount and place of waste discharge, water depths, etc., to calculate how the level of pollution will change with time and place. Without the computer, solution of the mathematical equations would be impossible. It is to be hoped that as skill in this sort of prediction increases, we will not discover too many instances of irreversible pollution.

~~~~~~

CHAPTER NINE

Desalination

IN THE United States today, 35 percent of the flow of rivers is withdrawn and 30 percent of that is consumed. In heavily populated areas even major streams may have all their water withdrawn and returned several times before they get to the sea. Water is withdrawn and returned with higher concentrations of dissolved material than when it was taken out. Sometimes the concentrations are increased simply because some of the water evaporates, raising the natural salt content of the remainder. Sometimes salts, of the same kind as those already in the water, are added as by-products of an industrial process, such as the sodium chloride added to the Detroit River by chemical plants along its banks. The sodium chloride is a residue of brines pumped out of salt deposits in the rocks far below the surface. Suspended and dissolved organic materials (and bacteria) may be added from sewage; exotic organic and inorganic chemicals may be dumped in

158

from a variety of industrial processes. The sodium content may go up from use of soap.

Water reuse, as concentration and contamination continue, becomes progressively more difficult and fewer and fewer customers can be found for many kinds of used water. Restorative water treatment of one kind or another has grown enormously in the past few years. It must continue to grow, and at an accelerating pace, because natural dilution of used water by pristine stream water diminishes as withdrawal increases. Water spoilage increases more rapidly than does rate of use.

Most water treatment now either adds something to water or substitutes one thing for another. Chlorine is added to kill bacteria, sodium is added to combat calcium hardness. For one reason or another there is the necessity for desalination—separation of water from the substances dissolved in it.

The magnitude of desalting operations today gives a hint of the scale of operations to come. Since the creation of the Office of Saline Water in 1952, the United States has become a leader in research and development of desalting techniques. The United States, with a 1970 production of about fifty-three million gallons a day (mgd) of desalted water from 322 plants, is second in production to the Middle East, where 74 plants produced nearly 63 mgd. Russia produced more than 37 mgd from only 7 installations, while in Europe, 88 plants produced nearly 30 mgd, and in the Caribbean islands more than 18 mgd came from 26 desalination plants. The size of the desalting installations ranges widely, from those with a production of only a few hundred gallons per day, to those, such as one in Rosarita, Mexico, with a capacity of 7.5 mgd. A plant scheduled to be built in Russia will eventually produce more than 31 mgd. At the present time, the largest plant in the United States, producing 2.6 mgd, is at Key West, Florida.

Although at the moment it is the dry areas that need desalted water, predictions are being made that even humid regions will turn more and more to desalination, in an effort to boost agriculture or to provide for large populations. Some crops need

159

continuous watering that even humid climates do not always ensure—desalination may be the answer. Special demands for fresh water in tourist resorts or in military installations may be met with desalination, a reliable source of supply and sometimes the only one.

The word that will be repeated many times in this chapter is cost. Can the cost of desalination be met? How much can, or will, be spent for water? How much is a gallon of water worth on the open market? Beyond the minimum required for direct human consumption, how much would be bid for *extra* water at a public auction?

The amounts that can or would be paid differ mostly in terms of water use. The extreme is found in special bottled waters. A gallon of demineralized water for a steam iron may retail for sixty cents. This is six hundred dollars per thousand gallons versus an average cost of drinking water of only six to fifteen cents per thousand gallons. For domestic uses like washing, cooking, and gardening, many people pay up to two cents per thousand pounds for water, or about two one-hundredths of a cent per pound. On this basis, a flush of the toilet costs about a tenth of a cent and a bath a half a cent. Industrialists will ordinarily pay about half as much for water as will a housewife, although for limited amounts of very pure water for special uses they may pay cents or tens of cents per pound.

Irrigation consumes so much water that it must have a cheap supply. Because of evaporation losses, even high value crops can be irrigated only with water that costs no more than ten cents or so a thousand gallons. Water cost is an important factor in controlling food prices. As the lowest current cost for desalted sea water used for irrigation is about a dollar per thousand gallons, the price of produce from fields irrigated with desalted sea water is raised tremendously. Production of fresh water from brackish water with 700 ppm dissolved salts costs about thirty cents per thousand gallons. This thirty cents is in striking contrast to the one cent that the average irrigationist can afford to pay.

These figures show that desalted water will be used widely for irrigation only if current costs can be reduced, except perhaps in a few places so isolated that desalination costs are offset by shipping costs for other water supplies. To assess the possibility of cost reduction in the future, the physical principles involved and the processes used must be examined.

One of the fundamental principles of nature is that all natural processes tend to "run down." If hot and cold water are mixed, the result is tepid water. If a concentrated salt solution and a dilute one are mixed, the final product is a uniform solution intermediate between the two. That differences in temperature always tend to be ironed out by mixing hot and cold is a matter of common knowledge, but what is not so obvious is that more energy can be obtained from a pail of hot water and a pail of cold water than from two pails of warm water. The mixing process, whether it involves temperature or concentration change, diminishes our ability to get energy from the substances involved. This decrease in the availability of energy is universal in the natural processes of the physical universe. Therefore, if sea water is to be separated into concentrated salty brine and nearly pure water, energy must be added in some form to "unmix" the sea water into these two parts.

The large amount of energy loss that occurs when salt dissolves in water is not obvious. If a teaspoon of salt is put into a pan of water used to cook vegetables, there is no discernible temperature change or other manifestation that anything drastic is occurring. The salt quietly dissolves and disappears. But if the pan were magnified enough to show single atoms it could be appreciated why it is so hard to get the salt back from the water. When sodium chloride dissolves, the sodium atoms and the chlorine atoms break apart in such a way as to leave each sodium atom with a positive electrical charge and each chlorine atom with an equal negative charge. These *ions* are immensely attractive to water molecules and become surrounded by a tightly bound sheaf of water molecules. The same kinds of processes occur when most other substances are dissolved in

161

water; limestone yields calcium ions and carbonate ions, each of which carries two electrical charges—positive for calcium and negative for carbonate.

When sodium chloride is dissolved in water, the volume of the resulting solution is less than the sum of the volume of the original water plus the volume of the solid sodium chloride. It is possible to dissolve solids in a full glass of water and end up with less than a full glass of solution. The water molecules bound to the ions take up less volume than the average molecule of liquid water. The effect of the ions is to "shrink" the water— literally to compress it. To reduce the volume of water by directly compressing it, rather than by adding salt, it would be necessary to have a squeezer that would apply tons of force to every square inch of water surface. The dissolving of salts to make brackish water or sea water is equivalent to subjecting the water to a terrific squeeze. The water that has been bound to the ions is almost impossible to remove, thus most processes of desalination involve separation of these ions, along with their bound-up water, from the remaining "free water."

To separate free water from the hydrated ions the free water can be evaporated and condensed from the solution into pure water vapor, or the free water can be frozen into pure ice, or the electrically charged ions can be pulled away from the free water with electricity, or the solution can be filtered through membranes that let only free water molecules through, or the free water can be combined with substances that make solid compounds separable from the hydrated ions.

But no matter how salts and water are separated, somewhat more energy is required to do this than was released when the salt originally dissolved, even for the theoretically perfect process. At today's electric power costs the lowest price per thousand gallons of fresh water from sea water by *any* process should be about five cents, an apparently irreducible minimum cost. The competition among the many processes is a competition to increase efficiency and to try to creep down toward that minimum cost from the current sixty-five cents to one dollar per thousand gallons.* If a 20 percent efficiency could be achieved—that is a cost of about twenty to twenty-five cents per thousand gallons—a revolution in water production by desalination of sea water would take place, especially if power costs are held down relative to other costs.

Desalination is not an invention of modern time. Long ago Aristotle described a method of evaporating sea water to produce drinking water. The Romans boiled water, collecting the vapor in sponges to be used as drinking water; Arabs in the eighth century A.D. distilled water. Thomas Jefferson demonstrated a method for shipboard distillation of sea water in the early part of the nineteenth century, and in 1869 the first patent for a desalination process was granted in England. The first large scale desalination plant came into use only in the 1930s on the Caribbean island of Aruba, where a plant with a capacity of six hundred and twenty-five thousand gallons a day was built.

Most of the fresh water produced from sea water today is by the oldest method known to man—distillation. Water is evaporated from a salty solution (leaving the salts behind) and is recovered by condensing it. In the simplest form, a dish containing salt water is covered with an inverted V-shaped glass and is placed in the sun; the water evaporates, condenses on the slanted walls of the glass cover, and runs down into "eavestroughs" along the edge of the dish. At first glance one would think that solar distillation would be an efficient, cheap method. Sunshine is free, the apparatus used has no moving parts and is fabricated from low cost, durable materials. Why then isn't every vacant lot in dry regions busy producing fresh water? Part of the answer is that it would indeed take every vacant lot to do the job.

* It is extremely difficult to give an accurate figure for the current cost in U.S. dollars for the desalination of a thousand gallons of water from seawater. Many different prices are quoted in many places, in various currencies, etc.; in some instances the costs are not for operating plants but are hopeful estimates resulting from pilot plant operations. At present, there seems to be no carefully documented instance of the actual production of desalinated water for less than one dollar per thousand gallons.

163

It is obvious that solar distillation has its greatest use in deserts where water is truly scarce and where there is empty land, sunshine, and high evaporation. But what about its potential as a panacea for world water shortages? About three feet of water evaporate each year from the sea surface—a tenth of an inch a day. As a world average, a box three feet on a side would yield about 3 quarts of distilled water per day—normal drinking requirements for two people. A family of four in suburban United States may well use 1,200 quarts of water every day, so they would have to have a distillation "farm" of 400 boxes, or 3,600 square feet of land surface, to take care of their daily requirements. There would probably be little room for lawns or patios or shrubbery, just space for the water supply system. For a community of 100 people requiring 25,000 gallons of water a day, about 5 acres of land would have to be devoted to solar stills. There would have to be storage tanks as well as stills, to guarantee a water supply during cloudy periods. Production in winter would be far less than during summer in most places. Solar distillation, at this time, is a poor method to satisfy global water requirements.

What is foreseen is that solar stills will have limited but significant application in the deserts of the world—to provide "existence" water in many places or even "affluent" water for people who can afford to live well and who also like deserts. For the person who wants to build a nice house in the desert and be independent of any local community, and to whom cost is not a serious consideration, a solar still can be quite a satisfactory answer to his water supply problem. All that is needed is access to water of almost any kind, brackish well water, a salty spring, or a river unfit for direct use. Stills can be bought complete for around a thousand dollars per unit of one hundred gallons a day capacity. Six percent interest on the initial capital investment would mean a minimum cost of sixteen cents per thousand gallons, to which must be added pumping costs, maintenance, taxes on the land required, and so on. Even so, for a desert hideaway, water costs would be an insignificant part of the total investment. At the other end of the scale, stills have performed miracles in places such as some of the small Greek islands where the inhabitants have been dependent on seasonal rain for their existence. In the worst years, water either ran out during drought months, or became so bad that everyone got sick. With solar stills, these small communities can at least have enough pure water for drinking at all times and are not under constant threat to their very existence. In some places enough water is generated to support a growing tourist trade.

Bright as it is, the sun doesn't have the heat intensity required to provide low-cost energy for distillation on a large scale. Coal, oil, or nuclear-fueled power plants take up much less space than solar stills in relation to the energy they produce, and can achieve a concentration of energy many times that of the sun. The distillers of water have done remarkable things. They have refined and refined the distillation process, utilizing every scrap of energy, until costs have been brought down below anything that was even hoped for a decade ago. Large plants with capacities measured in millions of gallons per day can now produce fresh water from sea water at eighty cents to one dollar a thousand gallons. There is still a big difference between that price and the ten or fifteen cent average cost of water at the tap, but there are many, many places where eighty cent water is welcome and cheap relative to the natural local supply.

What the distillers have done is to design apparatus that does not merely boil water, remove the steam, and condense it, while allowing the heat of condensation to be wasted. They know that because they start with water (salty to be sure) and end with water, they make a step toward achieving the magical five cents per thousand gallons to be achieved with 100 percent efficiency if they capture and use the heat released when the steam condenses. They still must worry, of course, about the cost of apparatus, water supply, water distribution, disposal of salt waste, and so on.

In one version of distillation, salt water is run into a vacuum tank where the reduced pressure causes part of the water to vaporize. The vapor is then condensed on the pipes that bring sea water into the tank so that the heat generated when the vapor condenses warms the water in the pipes, making evaporation easier. After part of the sea water has evaporated in the vacuum tank, it cools and is then led to a second vacuum tank, but on the way it is again warmed by the condensation of the vapor from the second tank, and so on. Water goes through the system without even being cooled very much. The more tanks there are, the closer the process comes to maximum efficiency. For obvious reasons the process is called *multistage distillation*. By using a partial vacuum to cause distillation, there is no need to heat the water to boiling and it can be desalted at ordinary temperatures.

The economics of multistage distillation are complicated. The more efficient the process, the more capital investment required for equipment, so that there is always a compromise between efficiency and power cost versus the plant cost.

In a variant of the multistage process, distillation takes place at the top of long columns; then the condensed water runs down along the outside of the column, giving its heat of condensation back to the saline water being processed within.

It is a tricky matter to obtain the minimum cost per thousand gallons by efficient operation of a given installation, or to predict what savings can be made in the future. For instance, in desalting sea water the cheapest fresh water from distillation is what is produced at the beginning of the process. As distillation continues, the remaining sea water becomes more concentrated and its proportion of "free" water diminishes as the proportion of water bound to the dissolved salts increases. If sea water is evaporated to a concentrated brine, the cost of distilling the very salty water at the end of the distillation system is much more than the cost of distilling the water at the start of the process. Again one cost factor balances another; tremendous volumes of sea water can be processed, distilling a small percentage from each volume to get high efficiency, but to do this pumping costs are high and a large plant is required. Also, for a large production of fresh water by distillation of a small percentage of sea water processed, the sea water supply must be huge. Even the problem of screening fish from the intake area may push the cost up considerably.

Desalination of sea water is big business today. There are about a thousand distillation plants with capacities of twenty-five thousand gallons a day or more scattered around the world and there are plans for facilities, linked to nuclear reactors, that will produce a billion gallons a day at a single site. Capacity today is small compared to such thinking for the future, but already annual world production is valued at several hundred million dollars.

Judging from the power requirements, reverse osmosis is the desalination method that has the greatest potential. There are more than a hundred reverse osmosis installations in operation, producing from 1,000 to 100,000 gallons a day. The process is an ingenious one and stems from a bright idea to force the natural process of osmosis to run backward. Long ago it was discovered that human cells, if put into a strong salt solution, shrank and shriveled; if put into pure water they swelled and burst. The membrane surrounding the cell contents permits water to move in and out, but it is reluctant to allow the electrically charged hydrated ions of the cell's salts to pass through. Cells contain several thousand parts per million dissolved salts. If a cell is put into sea water, which contains 35,000 ppm salts, the sea water has fewer "free" water molecules than does the cell fluid and the free water in the cell moves through the membrane into the sea water, in accord with the natural tendency of such systems to equalize. The cell fluid becomes concentrated, the amount of cell fluid diminishes, and the cell shrinks. For a cell in fresh water the process is reversed; water moves through the membrane into the cell, causing it to swell and eventually to rupture.

A cell membrane has to be very strong to resist the pressure created by water moving into the cell. The membrane surface is subjected to hundreds of pounds of pressure per square inch.

It is this *osmotic pressure* that is utilized by the cells of trees to solve their problem of getting nutrients from roots to leaves, a distance of perhaps a hundred feet to be traveled against gravity.

The analogue of a plant or animal cell can be made easily in the laboratory. If a concentrated salt solution in a cellophane bag is suspended in a container of fresh water, the bag swells as the water molecules move through the cellophane in their attempt to make the "free" water in the bag the same as that in the container.

The relation of osmosis to desalination was established when someone said, "Let's make a cell, but let's squeeze it so hard that even when it is put into nearly fresh water, water will move out of the cell fluid into the surrounding water. Let's overcome the osmotic pressure and make the system work the other way." Therefore, in theory, if a cell, or a cellophane bag, containing a salt solution could be squeezed hard enough, reverse osmosis should occur. Water would move out of the salt solution, leaving salt behind, making the remaining solution more concentrated, producing fresh water outside.

The important part of the whole idea was that the chief requirement was pressure, and pressure is cheap. Even though the pressures required are hundreds of pounds per square inch, which seems a formidable requirement, the apparatus is simple. In principle, all that is necessary is to put some brackish groundwater or some sea water into a cylinder with a membrane at the bottom end, put a piston down the cylinder from the top, and push. If the membrane is tight and strong enough to keep the electrically charged dissolved ions from migrating, what is squeezed out is pure water.

In practice there have been many problems. Membranes are made of thin cellophane-like material. It took a long time to find out how to mount them in the cylinder so that they would not break under pressure. Also, in this complicated modern world, pollution interferes and membranes become clogged if the feed water is not free of suspended material or if it contains oily substances. The membranes are not perfect filters, and the water coming through contains some dissolved salts (increasing in proportion to the salinity of the feed water). Nonetheless, the efficiency of the process is improving steadily as better membranes and new apparatus are designed. Because of the high salt leakage when concentrated solutions are used, most reverse osmosis systems have been applied to brackish waters containing a few thousand parts per million dissolved salts, rather than to sea water.

When salty water is frozen, the ice formed excludes dissolved salts. If part of a bucketful of sea water is frozen, the ice can be separated from the remaining brine and melted to produce fresh water. Initially, freezing is a more attractive method than distillation as a way of getting fresh water because the energy that has to be removed to cause freezing is less than the amount that has to be added to cause evaporation. In the end, the question is not fundamentally that of the energy requirements for freezing, but the degree to which the process can be made reversible. With the distillation process, a great deal of heat required for evaporation can be recovered by capturing the heat released during condensation. The question becomes one of higher efficiency and higher plant cost for freezing versus lower efficiency and lower plant cost for distillation.

At this date of writing, freezing has fallen behind distillation as a way to get fresh water, even though freezing was initially more attractive because the energy involved in making salt water into ice and brine is less than that to make water vapor and brine. In the not-too-distant future, freezing may forge ahead, however, as a major desalting process.

One of the most unusual of the desalination processes utilizes the formation of *clathrate* compounds. These are solids that form when gases are forced under pressure into pure or salt water. One ordinarily thinks of gases such as carbon dioxide or methane (marsh gas) as dissolving in water, but under pressure in cold water carbon dioxide unites with water to form solid crystals. These can be separated and melted to regenerate the carbon dioxide gas, leaving pure water behind. A sidelight on

169

clathrate formation comes from the Antarctic, where it was observed that gas bubbles in the ice cap are limited to shallow depths, disappearing in the deeper parts of the ice. A pretty piece of scientific detection showed that the gases in the bubbles reacted with the H_2O of the ice at low temperature and high pressure to form clathrates.

The clathrate process is probably the least investigated, and the least publicized, of the desalination processes, despite its many attractive features. Like the reverse osmosis process, it utilizes variations in pressure as the chief principle of its operation and therefore can be expected to have a high efficiency and low cost.

This review of desalination processes could be expanded greatly if it included all the variations on the basic themes: multistage flash distillation and long tube distillation; direct freezing and vacuum freezing; reverse osmosis and electrodialysis, in which membranes are used but an electric current pulls the dissolved ions from the water. Undoubtedly we will see plants that combine the different kinds of techniques, as well as the development of some new techniques unknown today. But however efficient they are, all have to face the reality of the minimum energy requirement.

Of all the techniques that have been suggested to solve water problems, desalination (especially of sea water) is the most controversial and the most exciting. On the one hand it offers the possibility of making the oceans usable, and thus opens up a virtually unlimited reservoir. On the other hand, even the most optimistic current estimates of the cost of fresh water reclaimed from the sea show a big gap between cost and what users are willing or able to pay.

It is our tendency to look at desalination on the side of the optimists; in it are all the kinds of problems that scientists and engineers have specialized in solving—problems in which the job was to translate the splitting of atoms into nuclear power plants, or to convert the propulsion systems of fireworks into a trip to the moon, or parlay the explosiveness of a mixture of air and gasoline into tens of millions of smoothly running automobiles. Research and development teamwork, in the parlance "R & D," is our great national talent. Salts are now removed from water on a limited scale; what is needed is further cost cutting by mass production with increased efficiency and ingenious methods and machinery.

By using nuclear energy, by using all kinds of tricks to utilize every bit of available heat, by building huge plants in the best situated areas—where the cost of obtaining sea water is least, where labor is cheap, where distribution of the purified water is to nearby consumers, and where the salty brine residue can be dumped directly into the sea—costs of twenty-five to thirty cents for a thousand gallons of fresh water can be hoped for.

At this price, who can afford the water? It has already been found that industry and the private citizen are willing to pay such amounts, if there is no other supply; but for agriculture, if we look at the general statistics of irrigation costs, no more than a few cents for a thousand gallons can be paid.

This economic picture begins to set up a framework for current use of fresh water from the sea. By careful planning, desert or near-desert areas close to the sea can be provided with adequate water at a cost that can be met, if all that is demanded is that the population survives on an industrial economy. The deserts near the sea can be populated if food can be imported, or if desalinated water can support subsistence farming. Deserts far from the sea have a serious problem in the disposal of the solid salts or concentrated brines that are the by-products of desalination. They also have high costs because feed water must be pumped to the desalination plant from underground sources. Another problem is the short life span of underground water sources.

The history of desert areas has been almost the reverse of the situation just described as being the most hopeful one. As a desert is populated, water is withdrawn from underground reservoirs and used for irrigation. Because deserts are popular if water is available, wherever they can be irrigated many people are likely to settle. Groundwater is quickly used up. The solution, so far, has been to bring new supplies in from afar, but costs are high and new supplies are dwindling.

Southwestern United States exhibits beautifully a classic example of this typical tragedy. Development of the southwest came at just the right time to show the desert story condensed into only two or three generations. When the pioneers moved west in the mid-1800s, they avoided the deserts as well as they could while traveling, and only a maverick few settled in them where an occasional spring seeped out, or where small reservoirs could be built to store water from the melting spring snows of the high mountains. Mines were developed and towns sprang into existence, dependent on water carried many miles by horse and wagon. The stream of settlers passed on and beyond the deserts, crossed the Sierra Nevada or the dry plains of eastern Oregon and Washington and homesteaded along the Pacific coast where water was quite sufficient for the first wave of settlers.

As the migration continued the land with enough water for crops was occupied and the second stage began—use of rivers flowing through the arid lands to irrigate the fields. The Rio Grande and the Colorado are two great exotic rivers, that is they flow from well-watered sources through deserts where they are no longer fed by rain or underground water seepage. They were exploited to water the desert soils and thus population followed the rivers. Where their valleys were broad a wide strip of green coursed through the desert brown.

Then the third stage set in as the pressure of population increased. The coastal district of southern California is naturally dry. Rainfall in the Los Angeles area is less than twenty inches yearly and the great V-shaped lowland that is now occupied by millions of people was almost a desert when the first settlers arrived. The only surface water came from streams flowing from the surrounding mountains, streams that rushed down their valleys in flood a few days or weeks of the year, then lay as dry gravel courses during the intervening months. But the soils were fertile and the settlement of the watered areas to the north gave a base of operation for developers. Small dams were built on the stream courses and some of the flood waters were saved for irrigation during the dry months. But more important was the discovery that the great thicknesses of gravel and sand on the flanks of the mountains that had washed down through eons of time were saturated with fresh water that could be pumped out cheaply. At the turn of the century the Los Angeles Basin became immensely attractive as a place to live. It was sunny, dry, and cheap. The population grew at such a pace that the statistics are really not believable. With growth, two things happened almost simultaneously. The reservoir of underground water became depleted and the groves of citrus trees were cut down. In a sense, the water demand was eased, because the new settlers used far less water in their homes than had been used for irrigation.

As time went on, the whole tremendous triangle of lowland known as the Los Angeles Basin became filled with homes and people. The orange and lemon trees and truck gardens disappeared beneath highways and homes. The underground water was nearly gone and the infrequently flowing streams were inadequate for the water demands. Even water brought by aqueduct across the mountains from the Owens River was not enough. The economy shifted from agricultural to industrial; hundreds and hundreds of light industries were built. But the new community needed food, whose production areas had been covered, and water, which had been exhausted. At this stage the City of Los Angeles had to begin dickering with the authorities controlling the Colorado River, miles and miles to the east.

This is the inevitable pattern of the southwest; irrigation from local water supply, overpopulation, exhaustion of water supply, development of a nonagricultural economy, importation of water, and then—or now, as far as Los Angeles and many other places are concerned—inadequacy of the supplies that can be commandeered from afar.

What is the next step? Perhaps it is to bring fresh water from still farther away (There is a plan to bring water from the Columbia River, a thousand miles to southern California). Another option is to have the people move away. This possibility seems remote in most cases. A final option is desalination. Not only Los Angeles, but the whole coast for a thousand miles to the south has the same water problem. It is next to the sea, fulfilling one major requirement for cheap desalination; it is a desirable area in which to live, judging from how avidly people settle in regions of pleasant temperature and high percentage of sunshine; and it is an area that could reasonably justify its existence if water were available for household and industrial use.

All this indicates that desalination, at costs of twenty-five or thirty cents for a thousand gallons of fresh water, might play an important role in those deserts that are near the sea and which can bloom as places to live and work. It is remarkable that if we lift our eyes from the west coast of the United States and look at the whole earth, we see that most of the deserts are next to the sea. But before looking at the potential of these areas, should desalination become widespread, consider the truly difficult problems of the deserts that have been heavily populated but are far from the ocean.

Los Angeles can still turn to the ocean, but what would be her fate if she were inland? Everything suggests that it would be rather grim. Tucson, Arizona, is a landlocked Los Angeles. Sitting beautifully in a magnificent lowland among the mountains and grown to some three hundred thousand people, its history is much the same as that of Los Angeles. What hope can be held out to Tucson? It is farther from any large fresh water source than is Los Angeles. As it has previously been pointed out (p. 63), it plans to develop every aspect of the local supplies. But in the end, Tucson, to survive and grow as it is growing, must have new fresh water sources. It does not look as if the Columbia River can be diverted there, and even optimum local development cannot help for long. Where, then, can Tucson turn? Desalination? The answer is a qualified and problem-fraught yes. Why not desalinate the ground water supplies? Or, with a limited total supply, can the water be recycled from salt to fresh, then made salt again by use, and again be recaptured?

Our guess is that the inland desert cities like Tucson will hang on about as they are, but that their potential for growth is small. It may well be that they will slowly decline, with the populace seeking areas equally desert in their character, but with the sea nearby to provide for the increasing water demands.

Returning to coasts and deserts and the potential of desalination, a fine general discussion of the problem has been published by Gale Young. We have drawn heavily on his article in *Science* entitled "Dry Lands and Desalted Water." * Young has worked out the current basic relationships between water and the major food crops of the world. Wheat, for example, has an average yield of 3 tons per acre, yields 1,480 calories per pound, and requires 91 gallons of water to grow (61 gallons of water for 1,000 kilocalories of food value). With a daily food intake of 2,500 kilocalories, a person living on wheat alone would indirectly use 150 gallons of water a day when he ate his 1.7 pounds of wheat. These facts make it possible to estimate how much can be paid for irrigation water relative to the cost of producing wheat. For instance, the desalination plants now selling water for $1 a thousand gallons would have to charge 15¢ for 150 gallons of irrigation water. The price paid to farmers for wheat is 4.7¢ per pound (8¢ for 1.7 pounds). The farmers would lose 7¢ after paying for irrigation water to raise that amount of wheat. Furthermore, the world average export price for wheat is only 2.9¢ a pound. The loss on growing wheat for the world market would be 10¢ on 1.7 pounds.

On the other hand, looking to the future, with multiple-function power plants that could also produce a billion gallons of desalted water a day, the price of water hopefully will be reduced to twenty-five cents a thousand gallons, or only three

* Gale Young, "Dry Lands and Desalted Water," *Science*, vol. 167, no. 3917 (1970), pp. 339–343.

and three-quarters cents for water to raise enough wheat to satisfy the daily food requirement of one person. This suggests that water costs and wheat prices might shift sufficiently to make irrigation of wheat with desalted water possible.

Rice requires several times as much water as wheat to grow. Although rice sells for about twice the price of wheat, water costs, even at twenty-five cents a thousand gallons, appear prohibitive for rice growers. There is hope, however, because recent experiments suggest that the cultivation of rice may be modified to require less water.

As Young points out, the picture for the grain crops is not hopeless and that for vegetables is much better. It may not be possible to maintain a complete range of agricultural products with desalted ocean water, but a desert community could take care of many of its food needs with specialized farming.

Young says, "Since much arid land lies relatively near the sea and the aggregate length of the coastal deserts nearly equals the circumference of the globe, desalination is a freshwater source of broad potential applicability when cheaper alternative sources are not available. Cities and industries can now spread widely along the ocean shore, bringing with them some agriculture for garden produce."

An attempt has been made here to show how complex it is to predict the future of a process like desalination, because of the many developments that might be influential. Power is required in enormous quantities; conventional fossil fuels already have many drawbacks, both in terms of supply and because of the pollution they create. Nuclear energy undoubtedly is required, but major increase of nuclear power also rests on many imponderables: Is there enough fuel for conventional reactors? Will "breeder reactors," which could vastly increase the nuclear power output, be developed? Will satisfactory methods of radioactive waste disposal be forthcoming? Or, in an entirely different direction, will the agricultural problems of the sterile tropical soils be solved so that areas of rapid future development will be the well-watered basins of the great rivers like the Amazon, to the neglect of the deserts? Or will there be so much attention to ending pollution that funds will not be forthcoming for development of desalination?

As was stated early in this chapter, we will bet on the engineers and scientists and predict that coastal deserts will see a rapid increase in population, their economy based on desalination, light industry, truck gardens, and tourism.

CHAPTER TEN

Pollution

WHO WANTS to return to the old days when food had a taste and water didn't and you couldn't even see the air you breathed? The realization of the deterioration of the environment, and of problems of population, food, power, and water, has been growing on us through a frightening total of isolated reports. Penguins accumulate DDT in Antarctica, the redwood forests are going fast, there are power blackouts in New York, oil fouls the Santa Barbara beaches, lead enters the atmosphere from gasoline, there is mercury in the Detroit River, radioactive wastes and nerve gas are dumped into the oceans. Every newspaper and news magazine brings another bad situation to our attention; there are crises everywhere. No matter what the real situation is, the constant bombardment creates a feeling of hysteria and despair.

The problems range from strictly local conditions, in which the cure is obvious, if not easy, like litter on the beaches, to worries about standing-room-only on the earth in a few generations. The problems are complexly interrelated, often in such subtle ways that solving one creates a whole chain of unanticipated others.

Maybe three of the most basic considerations are population, power, and the capacity of the environment. The gross influence of man on the environment can be measured as the sum of the energy used by all the people on earth. Five hundred years ago energy used could have been equated to the number of calories * of food consumed by each person daily. Today the energy used averages around sixty thousand calories per person. In the developed countries it is far more—on the order of one hundred and fifty thousand calories per person per day. Fifty times as much energy is consumed as is required for powering the human machinery alone. Where does it come from and where does it go? It comes from coal, oil, natural gas, and nuclear reactors, with a few percent contributed by hydroelectric power. It goes in a thousand ways but automobiles, planes, trains, and household uses account for about 30 percent of it.

With such a considerable percentage of energy used in automobiles and in the household, one lesson should be brought home to us—a great deal of the major additions to the atmosphere, the streams, and the oceans stem from individual demands for the good life. Everyone wants an automobile, or two, to take him where and when he wants to go, everyone wants stoves and washers and dryers and refrigerators and freezers and vacuum cleaners, and houses and cars that are warm in winter and cool in summer. Americans are the great exponents of the controlled and managed environment for the individual or the family, forgetting or not realizing that every ice cube made is coal or oil burned at a power plant, that dishwashers are responsible for sulfur dioxide in the air, that every time rugs are vacuum cleaned it is at the cost of more carbon dioxide released. The direct responsibility of the individual is most obvious in the use of automobiles. The driver of the car is running a power

* Large calories or kilocalories.

plant and can see the brown haze and fumes rising from the expressways.

The production of energy proliferates waste products. There is the first order contamination that results from the use of fossil fuels for power: carbon dioxide, carbon monoxide, sulfur dioxide and nitrogen compounds are among the contaminants. Nuclear reactors leave a residue of radioactive elements of many kinds. Then there is the second order contamination that springs from the use of the energy produced by the fossil fuels: paper products to be burned, chemical by-products to be disposed of, groundup garbage from macerators in sinks, old cars and motorcycles now littering the countryside in their graveyards.

Urbanization and technical advance account for much of the growing pollution. By 2000 nearly 95 percent of the United States population will live in urban areas. As the concentration of population increases so does the concentration of wastes. In 1954 approximately 190 billion gallons of waste-carrying waters were returned to streams, but by 2000 it is estimated the number will be 889. In 1966, 125 million tons of gases and dust were added to the atmosphere in the United States alone; estimates of additions for 2000 are as high as 600 million tons. The per capita production of solid wastes had grown from 2¾ pounds a day in 1920 to 4½ in 1965 and is climbing steadily. The range of pollution problems encompasses nearly every day to day human activity. Here we will emphasize those aspects intimately related to the web of water.

Throughout this book the water cycle has been emphasized: evaporation from the oceans, precipitation on the land, percolation through the soils to streams, return to the ocean. This is the basic circulation that keeps the world supplied with water. It is also the system that is joined by waste products. Many local circuits have already been overloaded devastatingly. Will a main fuse soon be blown? What are the most vulnerable parts of the cycle?

The complexities of the natural environment make it difficult to predict the consequences of interference with natural

180

systems. Ripples of reaction spread out from any pollutional stone cast into the environmental puddle. Many natural situations are remarkably stable and return to their original state even after a great disturbance; others are so delicately poised that they cascade to a worse state on being triggered by some apparently trivial event. Land areas dropped beneath the sea in the Alaska earthquake of 1964 have already been occupied by a typical balanced community of marine organisms. On the other hand, a single organism, like the lampreys that invaded the Great Lakes after the completion of the St. Lawrence Seaway and killed many game fish, may change the chances of survival of dozens of other species. The greatest worries about the building of a sea level canal across Panama are caused by lack of understanding of the results of mixing plants and animals from the two oceans. When they are abruptly thrown together as the waters of the two oceans mingle in a new canal, will the interactions be rapid and catastrophic? Will there be extinction of species far in excess of what would happen slowly if the organisms migrated naturally across natural barriers?

The question of the long-term results of man's current and accelerating environmental influence cannot be answered. Here the water cycle will be discussed with emphasis on the kinds of problems associated with different parts.

The surface of the sea is the primary source of rain and snow. Also, 30 percent of the salts dissolved in rivers come through the sea surface into the atmosphere, aided by bursting bubbles, and return to the continents in rain or snow. The amount of evaporation from the sea depends upon the amount of the sun's energy absorbed through the air-sea interface. Sunlight must come through the uppermost molecules of water to enable the microscopic plants of the ocean to carry out photosynthesis. The gases of the atmosphere and those dissolved in the ocean must exchange through the air-sea boundary. Thus any pollutant that changes the properties of water surfaces and is present in significant quantities has a potential importance.

Currently, of course, oil is a major worry. In the last fifty years, thousands of millions of tons of oil have been shipped by sea. Because of spillage, washing of ships' tanks, and shipwrecks, five to ten million tons of oil have been added to the sea. Recently, oil wells have been drilled on the continental shelves and oil has escaped from uncontrollable wells. At Santa Barbara, California, about a million gallons have escaped into the ocean since February 1970. Wild oil wells off New Orleans released six hundred and thirty thousand gallons of oil into the Gulf of Mexico in one month in 1970. A little further to the east, at Tampa Bay, Florida, a tanker lost ten thousand gallons of oil after a collision during the same month. Incidents like these are reported regularly from one part of the world or another; from San Francisco to the south coast of England there is hardly a beach in the world that has not been affected. Even the most remote islands have been defiled by the gelatinous blobs of oil.

Oil spreads on the sea surface. Light oils, like kerosene, evaporate almost completely into the atmosphere; heavier oils also lose their volatile components, but leave a thick residue behind. It is the loss of the volatiles that make the oil impossible to burn from the water surface after a short exposure to air. The presence of the residual oil layers in coastal areas is well known, especially for the deadly coatings acquired by birds and for the slowly hardening lumps and pellets of sticky residue that coat the sand and rocks of the shore. Much of the oil that comes ashore is deposited in the sand in shallow water. In this zone of wave action and good aeration it is usually decomposed within a few months. In the meantime, swimmers need a good supply of a strong solvent and boat owners are kept busy cleaning hulls. In the open sea the residue separates into two fractions. In one, tiny droplets of oil are formed that mix with seawater and become widely dispersed. In the other, tiny droplets of water get into the oil and change it into a thick, greasy mess. In time it is broken into chunks by the waves. Some chunks eventually sink, others float for months, presumably looking for a clean bathing beach on which to land. Fortunately there are many kinds of bacteria that attack the oil and degrade it into carbon dioxide and water. In high latitudes, because bacterial decomposition is extremely slow, the oil lumps last almost indefinitely.

No one seems to know with any certainty whether there is an organic residue from oil that joins the natural dissolved organic material of sea water. If there is, then it could have important consequences for the transfer of gases into and out of sea water at the surface and could conceivably have an effect on marine organisms. A thick scum of oil prevents light from reaching the plants that need it for photosynthesis. How *thin* a film could interfere with photosynthesis is unknown. It has been found that many marine creatures rely on trace amounts of chemicals for communication, much as the gypsy moth on land finds its mate by flying along a highly dispersed trail of an odoriferous sexual attractant. Degradation products of oil yield chemicals that might interfere with similar chemistry in the oceans and lead to great confusion among the organisms. Even without this problem the magnitude of oil pollution is great enough to cause worry about the critical zone between surface water and the atmosphere.

Another sensitive spot in the water cycle is the control of water vapor condensation in the atmosphere. Perfectly pure water vapor is almost impossible to condense. The formation of clouds depends on some form of nucleation of the water droplets. Dust is known to be important, but not in a clearly understood way. Dust in the atmosphere also absorbs solar radiation and helps to control the energy that reaches the earth, as well as that radiated back from the earth's surface. The particle content of smog is many times that of preindustrial air. The effect on solar radiation is evident from the wan reddish light that filters down on smoggy days. Smog over the United States is now so extensive that it can begin to influence climate, in its effects both on cloud and rain formation and on light absorption. The general result of pollution by particles is probably toward an overall cooling of the atmosphere. However, because of their role in affecting condensation and cloud formation, particles may cause obvious local climatic changes, but be of little consequence on a global basis.

Addition of carbon dioxide to the atmosphere from the burning of fossil fuels is of major worldwide importance. It is still difficult to say whether the 10 percent increase accomplished to date has had measurable climatic effects. More carbon dioxide should increase the amount of heat retained in the lower atmosphere. Heat radiating back to space from the earth's surface is partially adsorbed and retained by carbon dioxide; thus the atmosphere should be growing warmer. Because carbon dioxide and water are the basic ingredients for plant growth, a significant increase in carbon dioxide could increase the rate of growth of plants. Present uncertainty about the carbon dioxide situation raises some fascinating basic issues.

If it could be shown that the consequences of more and more industrial carbon dioxide additions would soften the climates of the polar regions without changing those of the tropics and otherwise appear to be quite beneficial, should an attempt be made to restore preindustrial levels or not? It is over issues of this kind that "preservationists" are separated from "managers." It is also the kind of situation that is a reminder that the major power controls of the sun-earth-atmosphere-ocean energy system are threatened. If altered, the system might possibly blow a few fuses in parts of the circuit where no trouble has been anticipated.

Because their tributaries branch throughout the countryside, rivers are susceptible to additions of every kind. They tend to be purest in their headwaters for several reasons: river sources are often in hilly or mountainous country with low population densities, their volumes are small, and their steep-walled valleys do not encourage the growth of industries. Where they are swift and turbulent, rivers have remarkable self-healing powers; air is mixed with the water in falls and rapids, and the dissolved oxygen decomposes organic wastes.

Where rivers pass through farm and pastureland their chief contaminants are animal wastes and fertilizers, sources of nitrates and phosphates. Much of the phosphate applied to the soil becomes fixed there, but the excess nitrate dissolves in soil water, percolates downward, and eventually wends its way to the streams. Because of its role as a plant nutrient, nitrate can have effects like those of phosphate in increasing the growth rate of aquatic plants and causing eutrophication. To date most such effects have been local. Major streams have not been markedly enriched in nitrate. The Mississippi River carries about 1 ppm nitrate; the upper limit for nitrate in drinking water is 45 ppm. Even the Colorado River with its 700–1,000 ppm total salt content and its heavy use for irrigation is not importantly polluted with nitrate. As demand for food increases, the number of livestock grows. Some hundred million cattle, sheep, and pigs add hundreds of millions of pounds of liquid and solid wastes to the landscape each day, but their contribution to stream pollution is relatively small.

It seems that the potential of the soil as a place to dispose of wastes has been much underrated and little investigated. The natural role of soil is to oxidize organic matter. In fact, under tropical conditions of high rainfall and temperature, organic materials are decomposed so rapidly at the soil surface that the forest litter hardly gets down into the soils at all. Soil particles have strong preferences for removing certain dissolved elements; they prefer potassium to sodium, for example. No one knows exactly what the controlling reactions are, but waters issuing from soils are always low in potassium, even if the soil particles are rich in that element. Some limited experiments using soils to store and filter radioactive wastes show that soil is generally effective in absorbing radioactive cesium and preventing its movement in ground water. A major future use of soils may be as natural waste water and garbage treatment plants. Ground up pea pods from a frozen food plant had been overwhelming a stream. As an experiment, they were disposed of by spraying them into the air as a sort of soup. When they fell to the forest floor below they were partly destroyed and partly incorporated into the soil.

Below their headwaters, streams become less turbulent and

self-renewing, and also tend to be the sites of towns and cities. Major pollution is by organic wastes—sewage, oil, garbage. All are biodegradable, that is bacteria decompose them. Oil in rivers causes a particularly difficult problem. As in the oceans, some of the oil tends to include water droplets and become a dense sticky mass. It then picks up sand or silt particles and sinks to the bottom. There it degrades slowly, depleting the oxygen in the bottom environment and creating gases such as hydrogen sulfide and methane. It may float back to the surface for a time, then return to the bottom. Organic sludges cause a similar problem. They settle on the bottom where they wipe out organisms—by covering them and so literally causing suffocation or by reducing the oxygen level. Once the organisms have been destroyed, they can return only after the sludge or oil has been oxidized away, but even then their return is gradual. Brandywine Creek in Delaware received a load of oil more than fifteen years ago and still will not support a normal bottom community of organisms.

Many chemical wastes, when mixed with water, cause a precipitate. Some mine wastes will thus yield a reddish brown iron oxide that fouls many streams. Although the precipitate itself may be neither toxic to organisms nor damaging to water composition, it can be harmful if it prevents light from reaching the plants or if it settles upon the bottom-dwelling organisms. Many solid wastes can act in the same way.

Just as the sea-atmosphere interface is a sensitive spot in the water cycle, so the contact between water and sediment is another. Much more is known about exchanges between sea and air than about communication between bottom sediments and the overlying water. The time necessary to clean up water pollution often depends upon bottom conditions. If Lake Erie has its water renewed, will it continue to be stagnant and polluted simply because the phosphorus that has accumulated in the bottom sediments will feed back into the Lake? If it does, plants could continue to flourish and through their decay maintain an oxygen deficiency in the Lake. How about the mercury in the Detroit River that has escaped from nearby chemical plants? If it has gone into the river sediment it may be a source of mercury for the river water for a long time to come.

The answers to many of these questions are not yet known. In general, heavy metals such as zinc or lead, when released in some dissolved form to a stream, quickly leave the water and accumulate in the bottom muds. Prospectors for these metals learned long ago that they could trace the origin of lead and zinc in a stream much better by analyzing the bottom sediments rather than the water.

Many pollutants are removed from solution by attaching to the surface of sand or mud particles. They may be returned to the water if they are replaced on the solid surfaces by other substances. Strontium, for example, is held much more firmly than zinc on most sediments. In fact, one of the worries about radioactive strontium is its tendency to replace other elements and so to be concentrated in soils or sediments. Bacterial activity can release from sediments some of the pollutants attached to sand or mud grains. Which substances remain in the sediment and which are subsequently released into the water depends upon general bottom conditions. One kind of behavior characterizes elements in the bottom of oxygenated lakes and rivers, another takes over when oxygen is deficient and eutrophication has begun.

Burrowing organisms that are nourished by absorbing organic materials from sediments are particularly susceptible to pollutants that cling to sediments. Thus clams and oysters are vulnerable to pollutants in the surface or upper layers of sediment.

Estuaries combine many of the features that make water susceptible to pollution. They are shallow, there is mixing of fresh and salt water, circulation tends to be slow. The organisms of an estuary are usually found in a delicately balanced succession from fresh water forms to true marine species. Any foreign influences such as toxic chemicals or large amounts of suspended materials can easily disrupt the niches that these organisms have found for themselves. Chesapeake Bay, Delaware Bay, the estu-

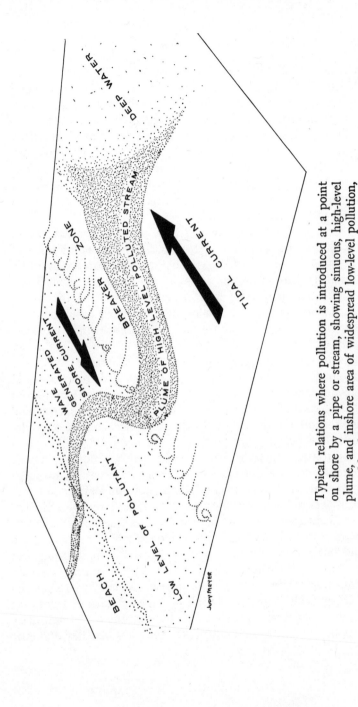

Typical relations where pollution is introduced at a point on shore by a pipe or stream, showing sinuous, high-level plume, and inshore area of widespread low-level pollution, caused by shifts in currents and waves.

ary of the Hudson—all are in a precarious position today. The Potomac River, below Washington, D.C., suffers from heavy additions of waste waters from Washington and from Alexandria, Virginia. The phosphorus content is high and plant growth is accelerating. The bays and coves within Chesapeake Bay, so charming when undefiled, are particularly easy to alter. Because of their protected position they are flushed only slowly by the tides.

Rivers, lakes, estuaries, and the nearshore areas of the oceans are being subjected to thermal pollution; large quantities of hot water are being discharged from power plants and industrial sites. Trillions of gallons of heated water are discharged into our waterways yearly. Thermal pollution threatens marine life by depleting the oxygen supply, by impairing the movement of fish so that they are unable to feed properly, etc. Thermal pollution can be controlled simply by cooling hot waters before discharging them. This is most often not done because it is cheaper to dump hot wastes than it is to cool them first. The problem of thermal wastes can be turned to good use—warm water can be used to encourage growth of certain fish and plants. Waste heat has been used successfully in fish farming to stimulate the growth of oysters and lobsters. Warm irrigation water can prevent frost damage or help to extend growing seasons by bringing premium-priced crops to an early market. Thermal discharges have been beneficial in keeping rivers from freezing in winter—a boon to the shipping business. Someone has even made a most ingenious suggestion that heat from power plants be used in Southern California to create ventilation that will disperse smog! Despite all these new ideas for making thermal pollution a positive influence, the picture is not always optimistic. For instance, the Columbia River and San Francisco Bay are both feeling the deleterious effects of hot water. In the Columbia, the increased heat threatens the life of salmon and steelhead, valuable commercial and sports fish, while the Bay is suffering from an oxygen deficiency.

These are some of the many results of pollution. Because each situation must be treated as a special case, it cannot be said that thermal wastes are always harmful nor that they are always acceptable. The difficult kinds of choices that must be made are not clear cut. Judgments will have to be made by balancing positive and negative. Probably the choice will not be open to decide between pollution or no pollution; undoubtedly it must be decided which waters will carry wastes and which will not, which rivers should provide good fishing and which should not. Economics will certainly enter the decisions. It may be too costly to restore every body of water to virgin condition.

The initial development of atomic reactors to produce atomic bombs for national defense and their proliferation as power sources has created some difficult problems. The reactions that produce fantastic amounts of energy from a small amount of material also produce a variety of radioactive by-products. These products have to be isolated from living organisms until their radioactivity has died away. Safe places must be found to keep them for the duration of their radioactivity, in some cases thousands or millions of years.

Everyone is subjected to a low level of natural radiation and man evolved successfully in its presence. Some of the potassium in our bodies is very slightly radioactive, as is some of the carbon. The ground itself contains traces of uranium, thorium, and radioactive potassium, all of which produce high energy radiation that is damaging in large doses. If the level of radiation is raised significantly above this "normal background" level, then the consequences are serious.

It has been hard to define the radiation levels that are dangerous. There is a vague suggestion that some environments are entirely uninfluenced by man's contribution but have a high enough natural level of radioactivity from rock and soil minerals to be dangerous to man. There are natural hot springs whose water cannot be drunk without risking severe radiation effects. The most perilous aspect of radiation is that at fairly low levels it can damage germ cells enough to cause abnormalities in offspring. At high levels it is directly lethal or results in burns and other bodily damage that sometimes never heal.

In trying to determine the levels of radioactivity that can be tolerated in the environment, many factors must be considered. Some of the products from atomic reactors are initially highly radioactive, so much so that exposure to them when they are first formed would produce such symptoms as burns and anemia, but which lose their radioactive character so fast that after a few days or months they are harmless, with no detectable activity remaining. These wastes are not a serious problem, for, unless by some grievous mistake they escape in large quantities directly from the reactor, they can be stored temporarily and then disposed of safely. Fortunately, almost all radioactive materials, except for their radiation, are essentially identical to ordinary chemical elements so that when their radioactivity has decayed, disposal of the remaining material is not hazardous.

There is a sort of transition group of radioactive materials that take a few years to decay. They are tougher to cope with, but they can be dumped into the ocean in sealed containers, for example, and their radioactivity is lost before the container can be corroded away by seawater. Even if a container broke immediately and the wastes got into the ocean, they would be diluted and would decay harmlessly. Some of these low level (rapidly decaying) wastes are buried where they can mingle with groundwater, but in places where the water moves so slowly that the radioactivity will have become harmless before the water trickles into any surface stream. It is not difficult to find locations where subsurface water requires many tens of years to percolate into the nearest stream.

So far there has been little general contamination of air, land, or sea by radioactive materials—if by contamination is meant the increase of radioactivity to levels known to be harmful to human beings. This apparently is true even in the food chain, where the concentration of many substances normally takes place. It is true that the levels of radiation of several elements are measurably higher than they were before nuclear fission—this is due chiefly to the residues of atmospheric testing of atomic bombs—but so far the development of nuclear energy for power has been a remarkably "clean" operation. Controlled discharges of low level wastes have been made into the ocean and the fate of the radioactive substances has been followed in detail by monitoring the water, bottom sediments, and marine organisms. It has been possible to mix the wastes with ocean water fast enough to prevent their accumulation in organisms and to permit the radioactivity to die harmlessly.

A way must be found to store the high level wastes, containing chiefly strontium and cesium, for the thousand years or so it will take to reduce their levels of activity to the point where they can be safely released to mix freely in the surface environment. At present these wastes are stored in steel tanks set into the ground and jacketed by concrete. The waste is a mixture of sludge and concentrated salt solution. Even though the strontium and cesium make up little of the total bulk of the waste, the mixture is still strongly radioactive.

All of these wastes from the United States, Great Britain, France, Israel, USSR, and China could probably go into a small lake basin, and if the strontium and cesium were separated out they would constitute a minute fraction of the total waste. A method needs to be developed to contain them for the thousand years that they must be kept out of the water cycle entirely. The pervasiveness of the web is clear when one begins to find out how this is done. Every plan for disposal must consider the possibility that within a thousand years the wastes will be mixed with water. As fossil fuel is replaced and supplemented by nuclear power, more of these wastes are produced and the disposal problem is becoming more and more severe.

The slow decay of strontium and cesium precludes dribbling them into the oceans. A batch of low level wastes can be added to the ocean and allowed to become harmless before another batch is added. High level wastes have such long lives that even small amounts put into the oceans at infrequent intervals could eventually build up high levels of radioactivity.

To put it mildly, nuclear power production is not without its waste problems. Whether the waste problem can be solved wholly or in part by modifying nuclear reactions is an open question, but one that is being investigated.

The system of disposal in the United States today is still that inherited from a sparsely populated country and the system has been continued despite an obvious proliferation of people, industry, and wastes. A small amount of waste in a large river creates few difficulties when the volume of water is great enough to dilute the waste. Disposal by dilution was an adequate system for many centuries. Today direct discharge of waste into rivers is overwhelming them rapidly. The degree of pollution of the Great Lakes, if measured by their added salt content, is directly correlated with the increase in the number of people living in the drainage area. Major rivers and lakes attract population and industry. In most of the world, population and industry have grown where they are best able to pollute streams and lakes! As noted before, animals discharge a greater volume of organic wastes than do people, but filtered through the soil and subject to chemical reactions there, animal wastes have not created a problem of the same magnitude as has the sewage disposal in most cities. The "out of sight, out of mind" concept that has dictated most systems of waste disposal has been held much too long. It has been observed that any space without an obvious purpose becomes waste-disposal space.

The water problem today is tied to numerous factors outside the water cycle. The size of population, the proliferation of power and use of energy impinge directly on the water cycle and are responsible for many of the problems. It should be clear that pollution can affect the water cycle at almost any point.

What will happen between now and the year 2000? The laissez-faire policy of yesterday has been abruptly changed, and a strong national effort to control the addition of wastes to the environment has begun. What had seemed to be a large number of local problems of different kinds suddenly have been recognized as regional and national problems, bewildering in their complexity but ultimately traceable to the growth of population, industry, and power production. If, thirty years from now, satisfactory water supplies, relative to demand, are available, the United States will have achieved a major triumph. The task would be difficult enough if demand remained constant, but the prediction is that it will be several times what it is now.

A major effort will have to be directed toward maintaining or augmenting stream flow, especially in the arid and semiarid parts of the country. The major threat to streams is water removal by increased evaporation as the result of increased irrigation, increase in the surface area of reservoirs, and increased use of fresh water for cooling in power plants and industry.

One of the possibilities of decreasing water loss from irrigation, in addition to the obvious one of decreasing acreage, is through underground irrigation pipes that leak water directly to the root systems of plants. The possibility also exists that over the next ten or fifteen years new varieties of crops will be bred that will be more efficient in their water use. More remote, but not impossible, would be the development of controls on water-release mechanisms of plants. Most plants do not require as much water as they actually use so this is an obvious target for investigation. At the extreme of management practices would be widespread control of vegetation of all kinds, in which water wasters would be eliminated from the landscape and replaced by frugal species.

It appears that the volume of water in reservoirs will continue to grow. Most reservoirs were originally constructed for power, for irrigation, or for flood control, but they have become increasingly popular for recreational purposes and are therefore vulnerable to pollution. Creation of artificial lakes increases evaporation by diverting water from rivers and spreading it over a larger surface. If lakes are to be used for recreation, it becomes necessary to hold a constant high water level, and this maximizes the area of water from which loss occurs. Whether safe and cheap methods to diminish evaporation from water surfaces can be developed is uncertain. There are still formidable barriers to the widespread use of chemicals to reduce evaporation.

The inevitable major increase in water use for cooling purposes probably will be eased by using sea water wherever possible and by building closed recirculatory systems where fresh

PEOPLE

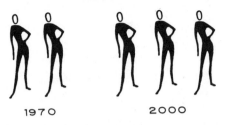

1970   2000

WATER PER PERSON

1970   2000

POWER PER GALLON

1970   2000

TOTAL POWER TO PROVIDE WATER

1970   2000

JUDY MEYER

Power requirements for water, with their attendant by-products, are pyramided by the combination of increasing population, increasing use per person, and increasing energy needed per gallon of supply.

water is needed. At any rate, with an estimated national requirement of 150,000 cubic feet per second of cooling water in the year 2000 (equal to about half the total flow of all the rivers of the Atlantic coast) the usual present practice of single use, direct flow-through cannot be continued.

If stream flow can be maintained, the accompanying step is a strong legislative program to curb water contamination. This part of the job is well under way. Current major developments are commonly in the direction of closed circuits of all kinds—circuits in which there is little net loss because water is continuously treated and recovered from an internal reservoir. Many factories now have built into them the equivalent of a sewage treatment plant and a desalination plant. Such systems require much energy; it may be that bringing chemical pollution under control will intensify the general problems of thermal pollution.

Almost everything that has been said points to more research, more technology, more legislation. There is not much hope for a "return to nature." Nature has already been changed so much by the activities of man that the course of the future is to interfere still more, with the hope that a careful, well-planned, and thoughtful program will have the magnitude and flexibility to cope continuously with all the unanticipated bad effects of the chain reactions that come from disturbing nature's balances.

A world in which man interfered with the rest of life no more than his biological share, that is to say by using just his required 2500 to 3000 calories of energy per day, might be defined as a truly natural world. But it seems impossible to keep him from building fires or smelting iron for his arrowheads. Now he is out of hand in his demands for quantities of energy the globe cannot continue to yield. To preserve his accomplishments—medicines for the sick, or books so that he can know his past, or protection against the elements—he must inevitably use much more than his share of energy. In the future of the race, the only safe source of that extra energy cannot be from the earth itself, it must come from the sun.

Bibliography

GENERAL REFERENCES

Books

Bardach, John. *Harvest of the Sea.* New York: Harper & Row, 1968.
Bloom, Arthur L. *The Surface of the Earth.* Englewood Cliffs, N.J.: Prentice Hall, 1969.
Briggs, Peter. *Water, The Vital Essence.* New York: Harper & Row, 1967.
Camp, Thomas R. *Water and Its Impurities.* New York: Reinhold, 1963.
Clark, John W. and Viessman, Warren, Jr. *Water Supply and Pollution Control.* Scranton: International Textbook Co., 1965.
Cook, J. Gordon. *The World of Water.* New York: Dial Press, 1957.
Fair, Gordon M. and Geyer, John C. *Water Supply and Waste-Water Disposal.* New York: John Wiley, 1954.
Frank, Bernard and Netboy, Anthony. *Water, Land and People.* New York: Alfred A. Knopf, 1950.
Furon, Raymond. *The Problem of Water, A World Study.* New York: American Elsevier, 1967.

Garrels, R. M. and Mackenzie, F. T. *Evolution of Sedimentary Rocks.* New York: W. W. Norton & Co., 1971.

Hardy, Alister C. *The Open Sea.* London: Collins, 1956.

Hirshleifer, Jack; De Haven, James C.; and Milliman, Jerome W. *Water Supply, Economics, Technology, and Policy.* Chicago: University of Chicago Press, 1960.

King, Thomson. *Water, Miracle of Nature.* New York: Macmillan, 1953.

Kuenen, P. H. *Realms of Water, Some Aspects of its Cycle in Nature.* New York: John Wiley, 1955.

Lewis, Alfred. *This Thirsty World; Water Supply and Problems Ahead.* New York: McGraw Hill, 1964.

Marx, Wesley. *The Frail Ocean.* New York: Sierra Club-Balantine Books, 1969.

Milne, Lorus and Margery. *Water and Life.* New York: Atheneum Press, 1964.

National Academy of Sciences. *Resources and Man.* Prepared by W. H. Freeman & Co., San Francisco: 1969.

Popkin, Roy. *Desalination, Water for the World's Future.* New York: Praeger, 1968.

Russell, F. S. and Yonge, C. M. *The Seas.* London: Frederick Warne, 1928.

Wolman, Abel. *Water, Health, and Society.* Bloomington: Indiana University Press, 1969.

Selected Periodicals

Chemical and Engineering News

Science

Scientific American

USEFUL SOURCES OF DATA

National Academy of Sciences-National Research Council, Publication 1000–B. *Water Resources, A Report to the Committee on Natural Resources,* by Abel Wolman. Washington, D.C.: National Academy of Sciences, 1962.

U.S. Geological Survey Circular 476. *Principal Lakes of the United States,* by Conrad Bue, Washington, D.C.: Government Printing Office, 1963.

Bibliography

U.S. Geological Survey Circular 643. *Reconnaissance of Selected Minor Elements in Surface Waters of the United States, October, 1970,* by W. H. Durum, J. D. Hem, and S. G. Heidel, Washington, D.C.: Government Printing Office, 1971.

U.S. Department of the Interior, Federal Water Pollution Control Administration. *Water Quality Criteria, Report of the National Technical Advisory Committee to the Secretary of the Interior.* Washington, D.C.: Government Printing Office, 1968.

U.S. Department of the Interior, Federal Water Pollution Control Administration Division of Pollution Surveillance. *Trace Metals in Waters of the United States,* by John F. Kopp and Robert C. Kroner. Washington, D.C.: Government Printing Office, 1969.

U.S. Department of the Interior, Office of Saline Water. *Saline Water Conversion Report.* Washington, D.C.: Government Printing Office, yearly.

U.S. Public Health Service, *Drinking Water Standards,* Washington, D.C.: Government Printing Office, 1962.

Index

200

Index

Brines, of Red Sea, 133–134
Bromine, in ocean, 123

Cadmium, 37
Caesar, and tides, 148
Calcium
  geologic cycling of, 94
  history in ocean of, 126–130
Calcium carbonate, carried by rivers, 128
Calcium toxemia, 30
Cape Cod, hurricane in, 152
Carbon dioxide
  in atmosphere, 184
  and climate, 184
Caspian Sea, 90
Cells
  environmental influences on, 26
  fluids in, 26
  membranes of, 26
  salt tolerance of, 27
  percentage of water in, 26
Challenger expedition, 119
Chemical pollution, of streams, 186
Chesapeake Bay, pollution in, 155, 189
Chlorination, 38
Chlorine, in ocean, 123
Cholera, 37
Chromium, 37
Clathrates, 169
Cleopatra's bath water, 53
Climate
  and carbon dioxide, 184
  continental variation of, 90
  and dust, 183
  and groundwater, 66
  and lakes, 113
Clouds
  formation of, 80, 81
  seeding of, 79, 80, 87
Coccoliths, 128
Coliform bacillus, 38
Colorado River, 96, 102
  discharge of, 92
  drainage area of, 92
  use of, 69
Columbia River, 97
  discharge of, 93
  drainage area of, 93
Condensation, 75
  and atmosphere, 183
  energy required for, 75–76
  example of, 77

explanation of, 77
  on impurities, 78
  problems of, 87
Congo River, 88–90
Continental drift, 21, 136
Continental shelves
  deposition on, 131
  economic importance of, 135
Continental slope, 131
Continents
  extreme temperatures of, 52
  origins of, 16, 17
Cooling, and evaporation, 76
Copper, effect in water, 36
Corals, 126–127
  of Bermuda, 127
  in geologic past, 127
  of Great Barrier Reef, 127
Cumulative poisons, 33

Dairy products, water requirements for, 44
Dams, 101–102
  economics of, 102
  effect on beaches of, 147
  lifetime of, 102
David Copperfield, 60
DDT, 33
Death Valley, 18
Deep sea mining, 135
Deep sea ooze, 127
Deep sea sediments, 132
  elements in, 133
Deltas, 100
Denudation, of rivers, 91
Desalination
  and agriculture and economics, 175–177
  assessment of, 170, 171, 176, 177
  costs of, 160–176 passim
  energy requirements for, 161
  history of, 163
  and irrigation, 160, 161
  methods of, 162
    clathrate, 169, 170
    distillation, 163–169
    freezing, 169
  plants for, 159
  of sea water, 167
Detroit River, 111
  pollution of, 158, 186–187
Diamonds, 135
Diamond Lens, 13
Dickens, Charles, 60

201

Index

Index

flow to ocean, 53
in Los Angeles, 66–67
minable amount of, 59
movement of, 66
pollution of, 65
renewal time of, 65
Texas use of, 67
total, 59
U.S. supply of, 65
Gulf of Mexico, 92

Hail, 82
Hard water, 41
Hawaii, tides in, 148
Heat, and molecular motion, 73
Hoover Dam, 102
Hubbert, M. K., 150
Humans
water deficiency in, 32
water tolerance of, 32
Humidity, 73
Hurricanes, 82–84
in Cape Cod, 152
cloud seeding to prevent, 83–84
damage from in U.S., 83
in Miami, 152
Hutton, James, 94
Hydration of salt, 162

Ice
International Patrol, 86
total, 85
vapor pressure of, 75
water in, 57
Ice Age, in western U.S., 67
Iceballs, as water source, 58
Icebergs, 86
as water source, 57–58
Industrial water, specifications for, 46–47
Industry
water consumption by, 55–56
water requirements of, 44–45
International Ice Patrol, 86
International Tsunami Warning System, 151
Iodine, and goiter, 35
Ions, definition of, 161
Iron
effect in water, 38
and manganese, effect in water, 36
Iron ores, 21
Irrigation

ancient, 42
in future, 194
by Russian river diversion, 89
and water consumption, 55
Irrigation water, 42–43
specifications for, 43
Isaacs, John, 57
Israel, water requirements of, 67–68

Jet trails, 79
Judson, Sheldon, 91

Kennedy, V. C., 97–99
Kramer, Calif., borax in, 115

Lakes
African, 112–113
and climate, 113
closed, 114
detergents in, 110
eutrophication of, 109–110
evaporation control of, 117
extinction of, 115–116
freezing of, 108
management of, 117–118
man-made, 116
in Minnesota, 115
open, 114
origin of, 104–105
oxygen-deficient, 109
oxygenated, 108–109
percentage of fresh water in, 64
playa, 114
Lake Winnemucca, 114
pollution clean-up of, 65
recreational use of, 117
total volume of, 105
waste disposal in, 193
and water supply, 117
Lake Baikal, 113
Lake Bonneville, 116
Lake Erie, 105–106
oxygen loss of, 112
pollution of, 111–112, 186
renewal time of, 112
water mining in, 58–59
Lake Huron, 105–106
Lake Mead, 102
Lake Michigan, 105–106
circulation of, 107–108
freezing of, 108
seiche on, 153
temperature of, 107–108
water exchange time of, 64

Index

Lake Nyasa, 112
Lake Ontario, 105–106
Lake Superior, 105–106
Lake Tanganyika, 112
Lake Titicaca, 113
Lake Winnemucca, 114
Lamb, Hubert, 89
Lampreys, 27
    in Great Lakes, 181
Land area of earth, 17
Land use, and rivers, 103
Lavas, deep ocean, 135
Lead
    pipes, 37
    poisoning, 37
Lena River, 88, 89
Life
    development of, 23–27
    first, 26
    and temperature, 22
Liquid, definition of, 71
Livestock
    and stream pollution, 185
    waste production by, 185
    water requirements of, 44
Loch Ness, 113
Los Angeles
    groundwater in, 66
    water problems of, 172–174

Magnesium, history of in ocean, 130
Manganese and iron, effect in water, 36
Manganese nodules, on ocean floor, 133
Mapping, of ocean floor, 120
Marianas Trench, 120
Mars, 17, 18
Mattole River, 97–98
Mercury poisoning, 33–35
Meteor Crater, 18
Micawber, Mr., 60
Microcosm, 20
Mid-Atlantic ridge, 136
Midwest, water supply in, 69
Miller, Stanley L., 24
Minimata Bay, 35
Mining
    deep sea, 135
    of water, 56–57
Minnesota, lake types in, 115
Mississippi River, 88, 89, 90, 100
    compared to Nile, 90

discharge of, 92
nitrate in, 185
Mississippi River system, 93–96
Missouri River, discharge of, 95
Molecular motion, and heat, 73
Moon, 17
Multistage distillation, 165–166
Municipal water
    consumption of, 55
    supplies of, 39

National Academy of Sciences Water Resources Committee, 105
Nearshore zone
    description of, 154
    pollution of, 188
    water residence time of, 154
New Orleans, 100–101
Nile delta, 100
Nile River, 90
    comparison to Mississippi, 90
Nitrate, 37
North Sea, gas in, 134

Ob River, 89
Ocean
    age of, 16
    ancient, 16, 21
    boron in, 123
    bromine in, 123
    calcium in, 126–130
    chlorine in, 123
    circulation in, 51–52, 122
    constancy through time of, 50–51
    depth
        greatest, 120
        in Marianas Trench, 120
        sonic measurement of, 120
    density of, 121
    economic potential of, 137
    energy received by, 51
    exploration of, 119–120
    floor, mapping of, 120
    fluorine in, 123
    heating and cooling of, 51–52
    light in, 121
    radioactive wastes in, 138–139
    rubidium in, 123
    salt in
        in geologic past, 125
        percentage of, 121
        value of, 124

204

Index

Index

Reservoirs (*continued*)
evaluation of, 60–61
future, 194
Reverse osmosis, 167
plant capacities for, 167
principle of, 167–168
Rio Grande River, 96
discharge of, 92
drainage area of, 92
Rip currents, 145
Rivers
administration of, 102–103
Atlantic slope discharge of, 92
chemistry of, 93
as function of time and discharge, 97–99
Missouri River, 95
North Atlantic slope, 94
Platte River, 95
and rocks, 94, 95
South Atlantic slope, 94
denudation of continents by, 90–
92
discharge of
major, 89
in southwestern U.S., 93
total, 88
dissolved load in, 90–91
dissolved salts in, 93
drainage area of, Nile and Mississippi compared, 90
erosion by
effect of man on, 91
effect of vegetation on, 92
flooding of, 101
future, 102–103
general properties of, 97
management of, 101
pollution of, 184–185
by oil, 186
reversal of Russian, 89–90
run-off of
from continents, 90
total U.S., 92
silting of, 99–100
suspended load in, 91
in U.S., 93–97
use of
in future, 103
and land, 103
waste disposal in, 193
water in
California supply of, 97
consumption of, 158

per capita supply of, 88
withdrawal of, 158
see also Streams
Rocks, oldest, 17
Root systems of plants, 44
Rubidium, in ocean, 123

Salinity, of ocean, 121
Salt
deposits of
in Michigan, 125
in New York, 125
energy of solution of, 161–162
hydration of ions in, 162
value in ocean of, 124
in water, 31
San Diego County, water requirements of, 58
Santa Barbara, oil pollution in, 134,
182
Sea
cliffs, 146
surface, and pollution, 181–183
Sea water
desalination of, 167
plant capacities for, 167
evaporation of, 124–126
near Andros Island, 125–126
salinity of, 35
Sediments
deep sea, 132
elements in, 133
water in, 137
diamonds in, 135
gas in, 134
oil in, 134
Seiche, 152
in Japan, 153
on Lake Michigan, 153
Selenium, 37
Shore zone, pollution in, 154
Sickness, caused by water, 37
Silica, physiologic effect of, 36
Snow, 84
effect on climate of, 85
Snow line, 85
Soil, and pollution, 185
Solar distillation, 163
cost of, 164
in deserts, 164
in Greek islands, 165
limitations of, 165
space requirements for, 164
water production by, 164

206

Index

Solar energy, 196
  and water cycle, 54
Solar radiation, and dust, 183
Solid, definition of, 71
Sonic measurement, of ocean floor, 120
Streams
  dissolved load in, 130
  flow of
    future, 194, 196
    in Georgia, 69
    to ocean, 53
  pollution of
    by chemicals, 186
    by livestock, 184–185
    by metals, 187
  sea salt in, 143
  suspended load in, 130
  U.S. use of, 54
  water storage in, 64
  see also Rivers
Submarine canyons, 131
Sulfur gas, in tidal flats, 149
Suspended load, in rivers, 91
Swell, definition of, 141

Tennessee Valley Authority, 101–102
Texas, groundwater use in, 67
Thermal pollution, 189
  of Columbia River, 189
  of San Francisco Bay, 189
Thomson, Charles Wyville, 119
Tidal flats, 149
  sulfur gas in, 149
Tidal waves, see Tsunamis
Tide pools, 149
Tides, 147
  in Bay of Fundy, 148
  and Caesar, 148
  in English Channel, 148
  in estuaries, 156
  in Hawaii, 148
  and invasion of Normandy, 148
  power of, 150
Titanic, 86
Tooth decay, and fluoride, 36
Tornadoes, 82
Toxic elements, 31
  combinations of, 32
  specifications for, 32–37
Trade winds, 122
  dust in, 132
Transportation, of water, 145

Trieste, 120
Tsunamis, 150
  at Georges Bank, 152
  in Hawaii, 151
  International Warning System, 151
  in Newfoundland, 152
Tucson, water problems in, 63, 174–175
Turbidity currents, 131

Underground water, see Groundwater
U.S.
  area of, 92
  interior drainage area of, 92
  total river runoff of, 92
U.S. Office of Saline Water, 159
U.S. Public Health Service, 31

Vapor pressure, 74
  of ice, 75
Vegetables, percentage of water in, 43
Venus, 18
  atmosphere of, 19
  comparison to earth, 18–19

Waikiki Beach, 60
Walsh, Don, 120
Waste
  disposal
    in lakes, 193
    in oceans, 139
    in rivers, 193
  production
    by livestock, 185
    solid, 180
  products, in water, 180
Water
  budget of U.S.
    current, 62
    future, 62
  chemistry of
    in ocean, 123
    in rivers, 93–97
  consumption of
    definition of, 54
    domestic, 56
    industrial, 55–56
    and irrigation, 55
    municipal, 55
    U.S., 55
  cooling of, in future, 194

207